Building Modern CLI Applications in Go

Develop next-level CLIs to improve user experience, increase platform usage, and maximize production

Marian Montagnino

BIRMINGHAM—MUMBAI

Building Modern CLI Applications in Go

Group Product Manager: Gebin George

Publishing Product Manager: Pooja Yadav

Senior Editor: Nisha Cleetus

Technical Editor: Pradeep Sahu

Copy Editor: Safis Editing

Project Coordinator: Manisha Singh

Proofreader: Safis Editing

Indexer: Hemangini Bari

Production Designer: Joshua Misquitta

Business Development Executive: Kriti Sharma

Developer Relations Marketing Executives: Rayyan Khan & Sonia Chauhan

Production reference: 2080323

Published by Packt Publishing Ltd.

Livery Place

35 Livery Street

Birmingham

B3 2PB, UK.

ISBN 978-1-80461-165-4

www.packtpub.com

Above all, I would like to express my gratitude to the Divine Intelligence, the source of all creation, for all the blessings and opportunities that helped me bring this book to life.

To my dear old dad, Dr. Joseph Montagnino, who always cheered me on in all things math and tech. Without you, I would not be the fearless learning machine I am today.

To my dearest mother, Lilia Banogon Montagnino, thank you for always seeing the potential in me and for being my guiding light throughout life. Your unwavering belief in me has given me the strength to face any challenge and pursue my dreams fearlessly.

To my dear sister, Lillian, your unwavering support and encouragement in self-discovery have allowed me to pursue my passion and find my true creative self.

To my nieces, Carine and Maple, and nephew, Cayden, thank you for the joy and playfulness you bring to my life.

To Julia and her late father, Marc Howe - you and the team at Cyber Warrior, Inc were my tech-savvy guild of wizards to guide and inspire me, and for that, I am forever grateful. Your mentorship and friendship paved the way for me, and I'll never forget it.

And to all my cherished friends and family - you all keep me grounded, laughing, and full of love. Thank you for being my support system and for making life so darn wonderful.

– Marian Montagnino

Foreword

On November 10, 2009, a historical announcement was made by Robert Griesemer, Rob Pike, Ken Thompson, Ian Taylor, Russ Cox, Jini Kim, and Adam Langley. They announced to the world a new experimental language called Go. It was on this day that the Go programming language was open sourced and available for the world to use. As we wind the clock forward 12 years to 2023, the language is currently ranked as the 12th most popular language in the world on the Tiobe index.

In 2016, the Go team released the results of their annual survey for the first time. One question in the survey asked people what they were writing in Go. Here are the top seven results of that question:

- **2,247 (63%)**: A runnable/interactive program (or CLI)

- **2,174 (60%)**: API/RPC services (returning non-HTML)

- **1,886 (52%)**: Web services (returning HTML)

- **1,583 (44%)**: Agents and daemons (e.g, monitoring)

- **1,417 (39%)**: Libraries or frameworks

- **1,209 (34%)**: Data processing (pipelines and aggregation)

- **1,120 (31%)**: Automation/scripts (e.g, deployment and configuration management)

At the time in 2016, CLI tooling was the number one use of Go. 8 years later in the 2022 survey, API/RPC services top the list at 73%, but a close second is CLI tooling at 60%. What might be more exciting is if you search the web for the best languages to use for building command-line tools, Go will always be at the top of any list.

If you have picked up this book because you need to write a command-line tool and you're thinking about using Go, or you've already chosen Go but need help on how to write the best CLI possible, you've picked up the right book. Marian is a seasoned software engineer with many years of experience writing CLI tooling for the companies and projects she has worked on. This book is what she wished she had when she was asked to write her first command-line tool.

I'm certain that you will learn what you need from this book and that the book will help jumpstart your knowledge into writing runnable and interactive programs in Go. You can feel confident that what Marian is teaching you will be idiomatic Go, following the best practices, design philosophies, and guidelines known today for CLI development and useability.

William Kennedy

Managing partner at Ardan Labs

Go trainer, writer, and speaker

Contributors

About the author

Marian Montagnino is a Senior Software Engineer at Netflix with over 20 years of experience. Since the early nineties, when her family first got a home computer, she has been using the terminal and command line applications to navigate through text-based systems. In 1995, she held her first job as a SysOp, or system operator, for Real of Mirage BBS in Fair Lawn, NJ. Her early years discovering technology inspired her to continue learning about computers. She received her Dual Computer Science and Mathematics of Operations Research BSc from Rensselaer Polytechnic Institute and her Applied Mathematics MSc from Stevens Institute of Technology.

To the exceptional members of the Go community, particularly Bill Kennedy, Natalie Pistunovich, Mat Ryer, Steven Francia, and Erik St .Martin, your creativity and innovation continue to inspire me towards greatness.

Finally, thank you to the wonderful team at Packt for their help and support throughout the process; I couldn't have done it without you!

About the reviewers

Wisnu Anggoro is a backend engineer with more than 10 years of experience in developing and maintaining distributed systems. He has expertise in Golang (Go Programming Language) and microservices architecture to solve complex software problems by implementing innovative solutions.

Tushar Sadhwani is a long-term Python and Go developer, open source contributor, author, and speaker. He is currently a Language Engineer at DeepSource, where he works on building linters and code analysis tools for Python. He has been an advocate of static type checking for over three years, and has been contributing to static code analysis tools over the past two.

Table of Contents

3

Building an Audio Metadata CLI   47

4

Popular Frameworks for Building CLIs   79

Part 2: The Ins and Outs of a CLI

5

Defining the Command-Line Process   101

6

Calling External Processes and Handling Errors and Timeouts   125

Part 3: Interactivity and Empathic Driven Design

8

9

The Empathic Side of Development   213

10

Interactivity with Prompts and Terminal Dashboards   239

Part 4: Building and Distributing for Different Platforms

11

12

13

Using Containers for Distribution   317

14

Publishing Your Go Binary as a Homebrew Formula with GoReleaser   343

Preface

If you're interested in taking your command-line interface (CLI) application development skills to the next level, Building Modern CLI Applications with Go is the book for you. This guide provides a comprehensive and hands-on approach to building CLI applications from scratch, using the popular Go programming language. Not only will you learn how to use frameworks such as Cobra and Termdash, but you'll also discover how to integrate empathetic approaches that prioritize human-first design. This book covers the entire development process, from compiling and distributing your application across multiple operating systems, to releasing it with GoReleaser and expanding your user base through Homebrew formulas. With clear explanations, practical examples, and insightful tips, this book will transform you into a skilled and creative CLI developer, able to create powerful, intuitive, and user-friendly applications that will delight your users.

Who this book is for

This book is aimed at intermediate Go developers who want to expand their skillset and develop powerful, user-friendly command-line interface applications.

What this book covers

Chapter 1, Understanding CLI Standards, The Command-Line Interface (CLI) was initially created as a way to interact with Operating Systems before graphical user interfaces (GUIs) were invented. Although the GUI and web have become more common, there has been a resurgence of CLI development in recent years, particularly as an additional offering alongside a company's API. In this chapter, you will learn about the history and anatomy of the CLI, the principles of UNIX, and why Go is a compelling language for building CLI applications.

Chapter 2, Structuring Go Code for CLI Applications, This chapter serves as a guide for those who are unsure of how to begin creating a new CLI application. It covers popular ways to structure code, the concept of domain-driven design, and provides an example of a real-world use case for an audio metadata CLI application. By the end of the chapter, readers will have the necessary skills to develop an application based on their specific use cases and requirements.

Chapter 3, Building an Audio Metadata CLI, This chapter provides hands-on learning by walking readers through building an audio metadata CLI's use cases from start to finish. The code is available online and can be explored independently or alongside the chapter. Additionally, readers are encouraged to use their imagination to consider alternative ways to implement commands.

Chapter 4, Popular Frameworks for Building CLIs, In this chapter, the most popular frameworks for developing modern CLI applications will be explored, with a focus on Cobra and its ability to quickly generate the scaffolding needed for a CLI application. Viper, which easily integrates with Cobra and provides extensive configuration options for applications, will also be discussed.

Chapter 5, Defining the Command-Line Process, This chapter delves deeper into the anatomy of a command-line application, breaking down the different types of input such as subcommands, arguments, and flags, as well as other inputs such as stdin, signals, and control characters. It also provides examples of processing data for each input type and how to return the result in a way that is easily interpreted by both humans and computers.

Chapter 6, Calling External Processes, Handling Errors and Timeouts, This chapter will teach you how to call external processes and handle errors, including timeouts, that may occur when interacting with other commands or API services. The os/exec package is discussed, which allows for the creation and running of commands with various options, such as retrieving data from standard output and standard error pipes. Additionally, the net/http package is explored for calling external API service endpoints, and the chapter concludes with strategies for capturing and handling errors that may arise.

Chapter 7, Developing for Different Platforms, One of the things that makes building a command line application powerful is the ability to easily create code that can run on different machines, regardless of their operating systems. The os, time, path, and runtime packages are great tools to help developers create platform-independent code. In this chapter, we explore the functions and methods in these packages with simple examples, and show how to specify code for operating systems using build tags. By the end, you'll feel more confident in your ability to write code that works across multiple platforms.

Chapter 8, Building for Humans Versus Machines, Developing your command-line application with your end user in mind is an important aspect to consider for better usability. In this chapter, we'll explore ways to build for humans and scripts, use ASCII art to increase information density, and the importance of consistency for better navigation across different commands and subcommands.

Chapter 9, The Empathic Side of Development, In this chapter, you will learn how to use empathy to develop a better command-line interface (CLI) by considering the output and errors written, providing empathetic documentation, and readily available help and support for users. By rewriting errors in a way that users can easily understand, providing detailed logging and help features like man pages, usage examples, and bug submission options, developers can create an empathetic CLI that meets the user's perspective and provides them with reassurance.

Chapter 10, Interactivity with prompts and Terminal Dashboards, This chapter will show you how to improve usability by adding interactivity to your command-line application using either prompts or terminal dashboards. By providing examples and step-by-step instructions for creating surveys and dashboards, this chapter will help you build a more engaging and user-friendly interface. However, it is important to disable interactivity when not outputting to a terminal.

Chapter 11, *Custom Builds and Testing CLI Commands*, To enhance the stability and scalability of a growing Go project, it is essential to incorporate build tags with Boolean logic to enable targeted builds and testing. This chapter demonstrates the use of build tags and testing through a real-world example, the audiofile CLI, and covers topics such as integrating levels, enabling profiling, and testing HTTP clients.

Chapter 12, *Cross Compilation Across Different Platforms*, This chapter explains cross-compilation in Go, including the different operating systems and architectures that Go can compile and how to determine which is needed. It covers topics such as manual compilation versus build automation tools, using GOOS and GOARCH, compiling for Linux, MacOS, and Windows, and scripting to compile for multiple platforms.

Chapter 13, *Using Containers for Distribution*, In this chapter, we'll dive into Docker containers and how they can benefit you when testing and sharing your CLI application. We'll begin with the basics and gradually move onto more complex topics, such as using containers for integration testing. Additionally, we'll weigh up the pros and cons of using Docker, helping you determine whether it's the right choice for you. By the end of the chapter, you'll be equipped to containerize your application, test it through Docker, and share it with others via Docker Hub.

Chapter 14, *Publishing your Go binary as a Homebrew Formula with GoReleaser*, In this chapter, you'll learn how to automate the release of a Go binary as a Homebrew formula using GoReleaser and GitHub Actions. GoReleaser simplifies the creation, testing, and distribution of Go binaries, and GitHub Actions is a CI/CD platform that automates software development workflows. By creating a Homebrew tap for your application, you can simplify the installation process for MacOS users and reach a larger audience.

To get the most out of this book

To get the most out of this book, you should have intermediate-level knowledge of Golang. The book assumes that you're familiar with Go's syntax, data types, control flow, and other basic concepts. It focuses on more advanced topics, such as creating and testing CLI applications, using external libraries, and building and distributing binaries. If you're new to Go, you may find the material challenging, but if you have prior experience, you'll be able to build on your existing knowledge and take your skills to the next level.

Software/hardware covered in the book	Operating system requirements
Go 1.19	Windows, macOS, or Linux
Cobra CLI	
Docker	
Docker Compose	
GoReleaser CLI	

Install Cobra CLI: `https://github.com/spf13/cobra-cli`

Install Docker Desktop: `https://www.docker.com/products/docker-desktop/`

Install the Docker Compose plugin: `https://docs.docker.com/compose/install/`

Install the GoReleaser tool at: `https://goreleaser.com/install/`

If you are using the digital version of this book, we advise you to type the code yourself or access the code from the book's GitHub repository (a link is available in the next section). Doing so will help you avoid any potential errors related to the copying and pasting of code.

Download the example code files

You can download the example code files for this book from GitHub at `https://github.com/PacktPublishing/Building-Modern-CLI-Applications-in-Go`. If there's an update to the code, it will be updated in the GitHub repository.

We also have other code bundles from our rich catalog of books and videos available at `https://github.com/PacktPublishing/`. Check them out!

Download the color images

We also provide a PDF file that has color images of the screenshots and diagrams used in this book. You can download it here: `https://packt.link/F4Fus`.

Conventions used

There are a number of text conventions used throughout this book.

`Code in text`: Indicates code words in text, database table names, folder names, filenames, file extensions, pathnames, dummy URLs, user input, and Twitter handles. Here is an example: "Mount the downloaded `WebStorm-10*.dmg` disk image file as another disk in your system."

A block of code is set as follows:

```
func init() {
    audioCmd.Flags().StringP("filename", "f", "", "audio
      file")
    uploadCmd.AddCommand(audioCmd)
}
```

When we wish to draw your attention to a particular part of a code block, the relevant lines or items are set in bold:

```
var (
    Filename = ""
)

func init() {
    uploadCmd.PersistentFlags().StringVarP(&Filename,
        "filename", "f", "", "file to upload")
    rootCmd.AddCommand(uploadCmd)
}
```

Any command-line input or output is written as follows:

```
cobra-cli add upload audio [-f|--filename]
    audio/beatdoctor.mp3
```

Bold: Indicates a new term, an important word, or words that you see onscreen. For instance, words in menus or dialog boxes appear in **bold**. Here is an example: "Select **System info** from the **Administration** panel."

> **Tips or important notes**
> Appear like this.

Get in touch

Feedback from our readers is always welcome.

General feedback: If you have questions about any aspect of this book, email us at customercare@packtpub.com and mention the book title in the subject of your message.

Errata: Although we have taken every care to ensure the accuracy of our content, mistakes do happen. If you have found a mistake in this book, we would be grateful if you would report this to us. Please visit www.packtpub.com/support/errata and fill in the form.

Piracy: If you come across any illegal copies of our works in any form on the internet, we would be grateful if you would provide us with the location address or website name. Please contact us at copyright@packt.com with a link to the material.

If you are interested in becoming an author: If there is a topic that you have expertise in and you are interested in either writing or contributing to a book, please visit authors.packtpub.com.

Share your thoughts

Once you've read *Building Modern CLI Applications in Go*, we'd love to hear your thoughts! Scan the QR code below to go straight to the Amazon review page for this book and share your feedback.

https://packt.link/r/1804611654

Your review is important to us and the tech community and will help us make sure we're delivering excellent quality content.

Download a free PDF copy of this book

Thanks for purchasing this book!

Do you like to read on the go but are unable to carry your print books everywhere?

Is your eBook purchase not compatible with the device of your choice?

Don't worry, now with every Packt book you get a DRM-free PDF version of that book at no cost.

Read anywhere, any place, on any device. Search, copy, and paste code from your favorite technical books directly into your application.

The perks don't stop there, you can get exclusive access to discounts, newsletters, and great free content in your inbox daily

Follow these simple steps to get the benefits:

1. Scan the QR code or visit the link below

https://packt.link/free-ebook/9781804611654

2. Submit your proof of purchase
3. That's it! We'll send your free PDF and other benefits to your email directly

Part 1:
Getting Started with a
Solid Foundation

This part covers the Command-Line Interface (CLI) and its resurgence in popularity. The history, anatomy, and design principles of the CLI are discussed, with a focus on UNIX's philosophy and the benefits of using Go to build a CLI. The guide offers a step-by-step approach to building a new application, including code structure, domain-driven design, and an example audio metadata CLI application. Hands-on learning is encouraged with the example audio metadata CLI application, and popular frameworks, such as Cobra and Viper, are explored to speed up the development process. Overall, this part provides a comprehensive overview of the CLI and its practical applications in modern programming, offering valuable guidance to developers looking to build efficient and effective CLI applications.

This part has the following chapters:

- *Chapter 1, Understanding CLI Standards*
- *Chapter 2, Structuring Go Code for CLI Applications*
- *Chapter 3, Building an Audio Metadata CLI*
- *Chapter 4, Popular Frameworks for Building CLIs*

1
Understanding CLI Standards

The **Command-Line Interface** (**CLI**) is a text-based interface for humans, and computer interaction was initially designed as a way of interacting with an **Operating System** (**OS**) before the desktop graphical interface was invented. The CLI, as we know it today, was in popular use in the 1960s until the graphical desktop interface was developed a decade later. However, although most computer users are used to the **graphical user interface** (**GUI**) and web, there's been a resurgence of CLI development circa 2017. Popular and new use cases for the retro CLI vary, but its most popular usage is as an additional offering alongside a company's API for increased platform usage.

In this chapter, you will learn about the comprehensive history of the CLI, what it is today, and a breakdown of its anatomy. You will learn about UNIX's philosophy and how following its principles will guide you toward the creation of a successful CLI.

By the end of this chapter, you'll have a deeper understanding of the CLI, how best to design and implement proven, time-tested standards, and why Go, which has become an increasingly popular language, has a compelling case for being the best language to build your CLI.

In this chapter, we are going to cover the following main topics:

- A brief introduction and history of the command line
- The philosophy of CLI development
- Modern CLI guidelines
- Go for CLIs

A brief introduction and history of the command line

The CLI is the result of the evolution of a much broader human-computer interaction, specifically communication and language processing. Let's begin the story with the creation of the first compiler, which took us from using punch cards to programming languages.

About the history

The first computer compiler was written by Grace Hopper. A compiler was able to translate written code to machine code and lifted the great burden off programmers of the time consumption of writing machine code manually. Grace Hopper also invented the COBOL programming language in 1959. In that era, punch cards were used for data processing applications or to control machinery. They would contain COBOL, Fortran, and Assembly code. The compiler and advancement of new programming languages eased the task of programming.

The same year, the microchip was invented by Jack Kilby and Robert Noyce. Much less expensive, small-scale computers were made possible and, finally, *human-in-the-loop*, a back-and-forth interaction between the human and computer, became feasible. Computers were now multitasking and time-sharing systems.

At this point, keyboards became the main method of interacting with computers. By the early 1960s, engineers had attached a **Cathode Ray Tube** (**CRT**) monitor to the **TeleTYpewriter** (**TTY**) machine. This combination of the CRT and TTY was called a *glass TTY* and marked the beginning of what we consider the modern monitor.

In 1966, the CRT and teletype machine, which combined the technologies of the electric telegraph, telephone, and typewriter, were about to merge with the final missing puzzle piece, the computer. The teletype computer interface was born. Users would type a command, hit the *Enter* key, and the computer would respond. These were called *command-line interfaces*!

There were so many more exciting developments that followed, from the invention of ASCII characters in 1963 to the internet in 1969, UNIX in 1971, and email in 1972. Lexical analysis parsers in 1975 played a major part in the development of programming languages. Text-based adventure games provided amusement for the tech savvy by 1977, and the beginnings of the GUI emerged in the 1970s.

A network of these computers would not have been possible if not for the evolution of the telephone. In 1964, the acoustic **modulator/demodulator** (**modem**) was used to transmit data between a telephone line and a computer. The acoustic modem brought us the **wide area network** (**WAN**), **local area network** (**LAN**), and the broadband we know of today. LAN parties peaked in the 1990s and carried on well into the early 2000s.

In 1978, the first public dial-up **bulletin board system** (**BBS**) was developed by Ward Christensen and Randy Suess, who also created the **computerized bulletin board system** (**CBBS**). With a modem, users could dial into servers running the CBBS software and connect via a terminal program. Throughout the 1980s, BBS's popularity grew to fantastic heights, and even in the mid-1990s, BBSs served the greater collective market compared to emerging online service providers such as CompuServe and **America Online** (**AOL**).

> **Note**
>
> This deeper understanding of the history of the CLI may give you a greater appreciation for what it is. The terminal is a bit of a time machine. The use of many UNIX and DOS commands feels like you're standing on the shoulders of giants, looking down at the long computing history beneath it.

Introducing the CLI

Based on the history of the CLI, it's clear to see that it is a text-based interface that allows communication from user to computer, computer to computer, and computer back to user. It requires the same specific instructions and clear language as the earlier machines it evolved from. Now, let's dig deeper into CLIs to learn about the different kinds, how they are generally structured, and how and why they are used.

Anatomy

For any CLI, regardless of the specific type, it's important to understand the anatomy of the commands themselves. Without a particular structure, the computer would not be able to properly parse and understand its instructions. The following is a simple example that we will use to distinguish the different components of a command:

```
~ cat -b transcript
```

In the context of this example, the UNIX command, `cat`, is used to view the contents of the file, `transcript`. The addition of the `-b` flag tells the command to print the line number next to non-empty output lines. We will go into each component of the command in detail in the following subsections.

The prompt

A symbol on the terminal indicates to the user that the computer is ready to receive a command. The preceding example shows ~ as the command prompt; however, this can differ depending on the OS.

The command

There are two types of commands:

- **Internal commands** are commands that are built into the OS shell, stored within internal memory, and execute faster. A few examples include folder and environment manipulation commands, such as `cd`, `date`, and `time` commands. They do not require a search of the PATH variable to find the executable.

- **External commands** are commands that are not built into the OS and are only available through software or applications installed by other parties. These commands are stored within secondary memory and do not execute as quickly as internal commands. A few examples include `ls` and `cat`. These are usually located in `/bin` or `/usr/bin` in UNIX and require a search of the PATH variable to find the executable.

The previous example uses `cat` as the external command.

Arguments and options

Commands usually take in parameters for input that consist of one or many arguments and/or options:

- Arguments are parameters that pass information to the command, for example, `mkdir test/`.

  In the preceding code snippet, `test/` is the input parameter to the `mkdir` command.

- Options are flags, or switches, that modify the operation of a command, for example `mkdir -p test/files/`.

  In the preceding example, `-p` is an option to make parent directories if needed.

In the example at the start of this section, `-b` is an optional flag, shorthand for `--number-nonblank`, which tells the command to print the line number next to non-empty lines, and the filename, `transcript`, is an argument passed into the command.

Whitespace

For the OS or application to properly parse these commands, arguments, and options, each is delimited by whitespace. Special attention must be paid to the fact that whitespaces may exist within the parameter itself. This can cause a bit of ambiguity for the command-line interpreter.

Take care to resolve this ambiguity by replacing spaces within parameters. In the following example, we replace the spaces with underscores:

```
cat Screen_Shot_2021-06-05_at_10.23.16_PM.png
```

You can also decide to put quotes around the parameter, as in the following example:

```
cat "Screen Shot 2021-06-05 at 10.23.16 PM.png"
```

Finally, resolve ambiguity by adding an escape character before each space, as in the following example:

```
cat Screen\ Shot\ 2021-06-05\ at\ 10.23.16\ PM.png
```

> **Note**
> Although whitespace is the most widely used delimiter, it is not universal.

Syntax and semantics

The CLI provides the language for communicating with the OS or application. Thus, like any language, to be properly interpreted, it requires syntax and semantics. The syntax is the grammar defined by the OS or the application vendor. Semantics define what operations are possible.

When you look at some command-line applications, you can see the language being used. Sometimes, the syntax differs between tools; I will go over the specifics later in this chapter, but, for example, `cat -b transcript` is a command we've looked at before. The command, `cat`, is a verb, the flag, `-b`, is an adjective, and `transcript` is a noun. This is the defined syntax of the `cat` UNIX command: verb, adjective, noun.

The semantics of the command are defined by what operations are possible. You can see this by viewing the options of, for example, the `cat` command, which are usually shown in the **usage** section of the help page, which is output when a user uses the application incorrectly.

Help pages

Because the CLI is entirely text-based and lacking in visual cues, its usage may be ambiguous or unknown. A **help page** is essential to every CLI. To view a list of valid parameters and options, users may run the command followed by the help option, typically `-help`, `--help`, or `-h`. The `-h` option is an example of an abbreviated shortcut for the help command.

There's a common syntax used in built-in help and man pages and following this standard will allow users familiar with the standard to easily use your CLI:

- Required parameters are typically represented within angled brackets, for example, `ping <hostname>`
- Optional parameters are represented within square brackets, for example, `mkdir [option] <dirname>`
- Ellipses represent repeated items, for example, `cp [option]... <source>... <directory>`
- Vertical bars represent a choice of items, for example, `netstat {-t | -u}`

Usage

The CLI was the first interface between the user and the OS used primarily for numerical computation, but in time, its usage has expanded in many more practical and fun ways.

Let us see some of the uses:

- **Editor MACroS (Emacs)**, one of the earliest forms of a text editor provided in UNIX, is a CLI in the form of a mini buffer. Commands and arguments are entered as a combination of key presses: either a *Ctrl* character plus a key or a key prefixed by a *Ctrl* character and the output displayed within another buffer.

- **Read-Evaluate-Print Loop (REPL)** is a Python interactive shell that offers a CLI, and according to its name can read, evaluate, print, and loop. It allows users a play environment to validate Python commands.

- **MajorMUD** and **Lunatix** are just a couple of popular games that were available on bulletin board systems. As soon as programmers could turn CLIs into play, they did, and while these games were entirely text-based, they were certainly not lacking in fun!

- Modern video games call their CLI a gaming console. From the console, mod developers can run commands to debug, cheat, or skip part of the game.

- Helper programs often take in parameters to launch a program in a particular way. For example, Microsoft Visual Code has a command-line option: `code <filename>`.

- Some CLIs are embedded into a web application, for example, web-based SSH programs.

- Companies such as AWS offer CLIs alongside their API as an additional way of interacting with their platform.

Last but not least, scripting has allowed engineers to take their CLIs to a more interactive level. Within a shell scripting language, programmers can script calls to the CLI and capture and manipulate output. The output of one command may also be passed as input into another. This makes the CLI a very powerful resource for developers.

Types

There are two main types of CLIs:

- **OS CLIs**
- **Application CLIs**

OS CLIs

OS CLIs are often provided alongside the OS. This kind of CLI is referred to as a **shell**. It is the command-line interpreter that sits a layer above the kernel interpreting and processing commands entered by the user and outputting results and a text-based method of interacting with the OS as an alternative to the graphical display.

Application CLIs

The second type of CLI allows interaction with a specific application running on the OS.

There are three main types of ways users may interact with an application's CLI:

- **Parameters**: They provide input to launch the application in a particular way
- **Interactive command-line sessions**: They are launched after the application as an independent and text-alternative method of control

- **Inter-process communication**: This allows users to stream or pipe data from the output of one program into another

GUI versus CLI example

Let's give a clear example of how the CLI can reign over the GUI in speed. Suppose we have a folder full of screenshots. The names of each contain a space and we'd like to rename these files to replace the whitespace with an underscore.

GUI

With a GUI, there'd be several manual steps for renaming a folder full of screenshots that contain whitespaces throughout the filename. Let's show these steps within macOS, or Darwin:

1. First, we'd need to open the folder containing all the screenshots:

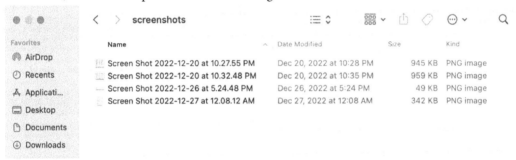

Figure 1.1 – Folder containing the screenshots where each filename contains numerous spaces

2. Second, we'd press the control button and left-click on a filename, then from the context menu that pops up, select the **Rename** option.

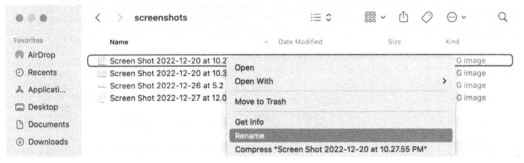

Figure 1.2 – From the context menu, click on the Rename option

3. Finally, manually replace each of the whitespaces with an underscore.

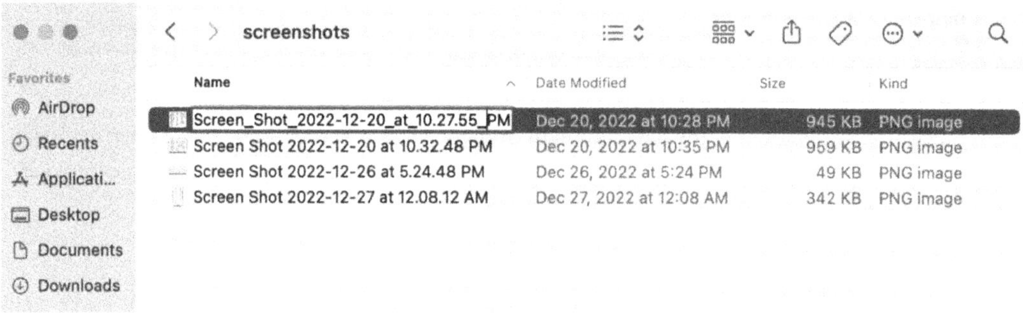

Figure 1.3 – Replaced whitespaces with underscores in filename

Repeat steps 1-3 for each file in the folder. We're lucky this folder only contains four screenshots. It can quickly get repetitive and tiresome with a folder of more files.

CLI

Let's see how much faster the CLI can be. Let's open the terminal and navigate to the folder with the files:

```
cd ~/Desktop/screenshots/
```

Let's view what currently exists in the folder by typing the ls command:

```
mmontagnino@Marians-MacBook-Pro screenshots % ls
Screen Shot 2022-12-20 at 10.27.55 PM.png
Screen Shot 2022-12-20 at 10.32.48 PM.png
Screen Shot 2022-12-26 at 5.24.48 PM.png
Screen Shot 2022-12-27 at 12.08.12 AM.png
```

Let's run a cleverly crafted command that loops through each file in the current directory and renames it (using mv) to translate the whitespace to an underscore:

```
for file in *; do mv "$file" `echo $file | tr ' ' '_'` ; done
```

Let's run the ls command again to see what's changed:

```
mmontagnino@Marians-MacBook-Pro screenshots % ls
Screen_Shot_2022-12-20_at_10.27.55_PM.png
Screen_Shot_2022-12-20_at_10.32.48_PM.png
Screen_Shot_2022-12-26_at_5.24.48_PM.png
Screen_Shot_2022-12-27_at_12.08.12_AM.png
```

Wow! We've just run a single command and the files are automatically renamed! This is just one example to show the power of CLIs and how much faster tasks can be executed compared to a GUI.

The comeback

The reason there's been a comeback of the CLI within recent years is because of these many benefits. The GUI can be resource-intensive, tedious when performing repetitive tasks, and sometimes slow.

The CLI, on the other end of the spectrum, is lightweight, scriptable, and fast. The advantages don't end there. The CLI might even offer commands and parameters that are not available, or are unthinkable, within the GUI. There's much to be admired and it's also a little mysterious.

I'm crushing a little on the CLI here! Jokes aside, to be fair to the attractive GUI, it has visual cues that allow the user to be self-guided. The all-too-mysterious CLI, on the other hand, requires help and man pages to understand its available parameters and options.

Though it may appear difficult to understand, once understood, the power of the CLI becomes apparent and inspiring.

The philosophy of CLI development

Philosophy plays a major role in the development of computer science. Throughout history, there have been many great contributions to computer science through philosophy, partially because many computer scientists were and are also philosophers. It is no surprise that each OS has its own distinct philosophy.

Windows, for example, hardcodes most of its intelligence within the program or OS, assuming users' ignorance and limiting their flexibility. Although the barrier to understanding is lower, users interact with the program without understanding how it works.

The developers of **UNIX** had an opposing philosophy: provide the user with almost limitless possibilities to empower them. Although the learning curve is steep, much more can be developed within an environment that doesn't shield its users from the complexity of freedom.

There have been many books written about UNIX's philosophy and implementing it in real life is an art form. I am sure, therefore, many people view coding as a craft. Although there are many other philosophies to review, the focus in this section will be on UNIX's philosophy.

The legendary designers of the Go programming language, Ken Thompson, Robert Griesemer, and Rob Pike, share a long history with UNIX, and it feels fitting to discuss the philosophy within the context of its creators since Go was built around it.

UNIX's philosophy advocates for simple and modular designs that are both extensible and composable. The basis is that the relationships between numerous small programs are more powerful than the programs themselves. For example, many UNIX programs handle simple tasks in isolation, but when combined, these simple tools can be orchestrated in a very powerful way.

Checklist for a successful CLI

The following are some principles inspired by this UNIX philosophy that when followed will help create a successful CLI:

- **Building a modular program**

  Design your CLI with standardization in mind to ensure it can be easily composed with other applications. Specifically utilizing standard in and out, standardized errors, signals, and exit codes helps to build a program that is both modular and easily composable. Composability can be handled simply with pipes and shell scripts, but there are also programming languages that can help piece programs together. **Continuous Integration/Continuous Delivery (CI/CD)**, orchestration, and configuration management tools are often built on top of command-line execution and scripts to automate code integration or deployment or to configure machines. Consider the data output from your program and how easily composable it is. The best options are plain text or JSON when structure is needed.

- **Building for humans first**

  The first CLI commands were written with the assumption that they'd only be used by other programs. This is no longer the case, and so programs should be built with humans first in mind.

  Conversation will be the main method of human-computer interaction. Imagine the natural flow of human conversation and how that concept can be applied to help a user who has misunderstood the program design. In natural language, your program can suggest possible corrections, the current state in a multi-step process, and request confirmation before continuing to do something risky. In the best-case scenario, your user has had a pleasant experience with your CLI, feeling empowered to discover operations and receiving assistance when needed. In the worst-case scenario, your user feels ignored and frustrated with no help in sight. Don't be that CLI!

  Finally, write readable code so other developers can easily maintain your program in the future.

- **Separating interfaces from engines and policies from mechanisms**

  Decoupling these allows different applications to use the same engine through interfaces or use the same mechanism with different policies.

- **Keeping it simple**

  Only add complexity when it's necessary. When complexity does occur, fold it into the data instead of the logic. Where usability is not compromised, use existing patterns.

- **Staying small**

  Don't write a big program unless there's no other way.

- **Being transparent**

  Be as transparent as possible so users can understand how to use the program and what's going on. Transparent programs have comprehensive help texts and provide lots of examples allowing users to easily discover the parameters and options they need and have the confidence to execute them. The GUI certainly has a leg up in terms of transparency and visibility; however, we can learn from it and see what can be incorporated to make the CLI easier to learn and use. Users resorting to Google or Stack Overflow is an anti-pattern here.

- **Being robust**

  Robustness is the result of the former principle: transparency and simplicity. The program should work in a way that the user expects, and when errors occur, explain what is happening clearly with suggestions for resolution. Immediately printing stack traces or not informing the user with a clear and immediate response leaves the user feeling like they are on shaky ground.

- **No surprises**

  Keep your program intuitive by building on top of a user's existing knowledge. For example, a logical operator such as + should always mean addition and - should always mean subtraction. Make your program intuitive by staying consistent with pre-existing knowledge and patterns of behavior.

- **Being succinct**

  Don't print output unnecessarily and don't be completely silent, leaving the user to wonder what's going on. There's a balance in communication required to say exactly what needs to be said; no more, no less. Too much is a large block of verbose text that forces the user to parse through it to find useful information. Too little is when the command prompt hangs in silence leaving the user to assume a state about the program.

- **Failing noisily**

  Repair what can be repaired, and when the program fails, fail noisily and as soon as possible. This will prevent incorrect output from corrupting other programs depending on it.

- **Saving your time**

  Build code to save developers' time as opposed to the machine's time, which is relatively cheap these days. Also, write programs that generate programs. It's much faster and less error-prone for computers to generate code over hand-hacking.

- **Building a prototype first, then optimizing**

  Sometimes, programmers spend too much time optimizing early on for marginal gains. First, get it working, and then polish it.

- **Building flexible programs**

  Programs may be used in ways the developers did not intend. Therefore, making the design flexible and open will allow the program to be used in ways unintended.

- **Designing for extensibility**

  Extend the lifespan of your program by allowing protocols to be extensible.

- **Being a good CLI citizen**

  Bring empathy into the design and peacefully coexist with the rest of the CLI ecosystem.

The philosophy directly influences the guidelines for creating a CLI. In the next section, you will clearly see the link to satisfy the philosophy tenets discussed, and if anything, following the guidelines will increase the odds of creating a successful CLI.

The guidelines

These guidelines have been formulated since the first CLI and have continued to evolve through the many years of developer and user experience. Following these guidelines will increase your chances of CLI success; however, there may be times when you decide to go your own way and follow an anti-pattern. There could be many reasons to choose an unconventional route. Remember that these are just guidelines and there are no hard and fast rules.

For life, and building CLIs, to be fun, we must allow a little chaos and the freedom necessary to be creative.

Name

The **name** of the CLI holds significant weight as the name may convey symbolic ideas beyond the initial intention. People do not like to think more than necessary, so it's best to choose a name that is simple, memorable, and easy to pronounce. It's amazing how many CLI program names have been chosen so arbitrarily without much thought.

There have been studies that support the linguistic Heisenberg principle: *labeling a concept changes how people perceive it.*

Hence, keep it short and easy to type. Use entirely lowercase variables in the name and only use dashes when absolutely necessary.

Some of my favorite application names are clear in the way that they plainly describe the application's purpose in a creative manner. For example, **Homebrew** is a package manager for installing applications on macOS. A brew is a concoction of various ingredients, like a recipe, or in this particular case, a formula, to describe how to install an application. Another great example is `imagemagick`, a command-line application that lets you read, process, or create images magically! Truly, as Arthur C. Clark writes, "*Any sufficiently advanced technology is indistinguishable from magic.*" Other internal commands we are familiar with are mkdir, for make directory, rm, for remove, and mv, for move. Their popularity is partially a result of the transparent nature of their names, rendering them nearly unforgettable.

Help and documentation

One of the tenets of the UNIX philosophy is transparency, possible mainly through the help and documentation present within the CLI. For new users of the CLI that are in discovery mode, the help and documentation are one of the first sections they will visit. There are a few guidelines to make the help and documentation more easily accessible to the user.

Help

It is a good practice to display help by default when just the command name is entered or with either the -h or -help flag. When you display the help text, make sure it's formatted and concise with the most frequently used arguments and flag options at the top. Offer usage examples, and if a user misuses the CLI, the program can guess what the user tried to attempt, providing suggestions and next steps.

Documentation

Provide either man pages or terminal-based or web-based documentation, which can provide additional examples of usage. These types of documentation may be linked from the help page as an extension of the resources for gaining an understanding of how the CLI works.

Support

Oftentimes, users will have suggestions or questions on how to use the CLI. Providing a support path for feedback and questions will allow users to give the CLI designer a new perspective on the usage of their CLI. When collecting analytics, be transparent and don't collect users' address, phone, or usage data without consent.

Input

There are several ways a CLI retrieves input, but mainly through arguments, flags, and subcommands. There is a general preference for using flags over arguments and making the default the right thing for most users.

Flags

The guideline for flags is that ideally, there exists a full-length version for all flags. For example, -h has --help. Only use -, a single dash, or shorthand notation for commonly used flags and use standard names where there is one.

The following is a list of some standard flags that already exist:

Flag	Usage
`-a, --all`	All
`-d, –debug`	Debug
`-f, --force`	Force
`--json`	Display JSON output
`-h, --help`	Help
`--no-input`	Disable prompt and interactivity
`-o, --output`	Output file
`-p, --port`	Port
`-q, --quiet`	Quiet mode
`-u, --user`	User
`--version`	Version
`-v`	Version or verbose
`-d`	Verbose

Table 1.1: Standard flags

Arguments

Multiple arguments are fine for simple actions taken on several files. For example, the `rm` command runs against more than one file. Although, if there exist two or more arguments for different things, you might need to rethink the structure of your command and choose a flag option over an additional argument.

Subcommands

The guidelines for subcommands are that they remain consistent and unambiguous. Be consistent with the structure of subcommands; either *noun-verb* or *verb-noun* order works, but *noun-verb* is much more common. Sometimes, a program offers ambiguous subcommands, such as `apt update` versus `apt upgrade`, which causes many, including myself, confusion. Try to avoid this!

Validate the user's input early, and if it's invalid, fail early before anything bad happens. Later in this book, we will guide you through using Cobra, a popular and highly recommended command-line parser for Go, to validate user input.

Output

Because CLIs are built for humans and machines, we need to consider that output must be easily consumed by both. I will break down guidelines for both `stdout` and `stderr` streams for both humans and machines. Standard output, `stdout`, is the default file descriptor where a process can write output, and standard error, `stderr`, is the default file descriptor where a process can write error messages:

- **stdout**

  A guideline for standard output for humans is to make the responses clear, brief, and comprehensible for the user. Utilize ASCII art, symbols, emojis, and color to improve information density. Finally, consider simple machine-readable output where usability is not impacted.

  A guideline for standard output for machines is to extract any extraneous substance from the above response so that it is simply formatted machine-readable text to be piped into another command. When simple machine-readable text is not output by default, utilize the `-q` flag to suppress non-essential output and `--plain` to display machine-readable text. Disable color with the `--no-color` option, by setting the `NO_COLOR` environment variable, or a custom `MYAPP_NO_COLOR` environment variable specific to your program.

  Additionally, don't use animations in `stdout` because it's not an interactive terminal.

- **stderr**

  Things can go wrong during the execution of a command, but it doesn't have to feel like a catastrophic event. Sometimes loud full stack traces are the response to a command failure, and that can make the heart skip a beat. Catch errors and gracefully respond to the user with rewritten error messages that can offer a clear understanding of what happened and suggestions for the next steps. Make sure there's no irrelevant or noisy output, considering we want it to be easy to understand the error. Also, provide users with additional debug and traceback information and an option to submit bugs. Non-error messages should not go to `stderr`, and debug and warning messages should go to `stdout` instead.

> **Note**
>
> As for general guidelines for CLI output, return a zero exit code on success and a non-zero code that the machine can interpret as not just a failure but even a particular type of failure on which to take further action.

Configuration

Users may configure their CLI by using flags, environment variables, and files to determine how specifically to invoke the command and stabilize it across different users and environments:

- **Flags and environment variables**

  By using flags or environment variables, users may configure how to run a command.

  Examples include the following:

  - A specified level of debug output

  - Dry run commands

Alternatively, they can be used to configure between different environments.

Examples include the following:

- Providing a non-default path to files required for the program to execute

- Specifying the type of output (text versus JSON)

- Specifying an HTTP proxy server to route requests through

When using environment variables in configuration, set names appropriately, using a combination of all uppercase text, numbers, and underscores, and take the time to ensure you are not using the name of an already-existing environment variable.

- **XDG Spec**

  Configure stability across multiple environments by following the XDG Spec (X Desktop Group, `freedesktop.org`), which specifies the location for base directories where configuration files may be located. This spec is supported by many popular tools, such as Yarn, Emacs, and tmux, to name a few.

Security

Do not store secrets and passwords in environment variables or pass them in via an argument or flag. Instead, store them in a file and use the `--password-file` argument to allow the secret to be passed in discretely.

Open source community

Once your CLI is complete and ready to be distributed, there are several guidelines to follow. If possible, distribute it within a single binary targeted to a user's specific platform and architecture. If the user no longer wants or needs your program, make sure it's easy to uninstall too!

Since you'll be writing your CLI in Go, it would be great to encourage contributions to your program. You may offer a contribution guideline doc that guides users toward commit syntax, code quality, required tests, and other standards. You could also choose to allow users to extend the CLI by writing plugins that can work with your CLI, break up functionality across more modular components, and increase composability.

Software lifespan and robustness

To ensure your CLI will continue to work well in the future, there are a few guidelines to follow specific to robustness to make sure your program has a long lifespan:

- **Future-proofing**

  When you make any changes to your CLI, over time, it's best to make these changes additive, but if not, warn your users of the change. Changing human-readable output is usually fine, but it's best to keep the machine-readable output stable. Consider how external dependencies may cut short your program's lifespan and think of ways to make your application stable amidst external dependency failures.

- **Robustness**

  For a CLI to achieve maximum robustness, the CLI must be designed with full transparency. The program needs to feel responsible for the user; so, show progress if something takes a long time and don't let the program hang. Make programs timeout when they are taking a long time. When the user inputs a command, validate the usage immediately, giving clear feedback when there's apparent misuse. When there's a failure due to some transient reason, exit the program immediately upon failure or interruption. When the program is invoked again, it should pick up immediately where it left off.

- **Empathy**

  Adding some thoughtful detail to the command-line design will create a pleasant experience for the user. CLIs should be fun to use! Even when things go wrong, with a supportive design, the users can feel encouraged on their pathway to successfully using the CLI. The modern CLI philosophy and guidelines reflect a level of empathy toward humans already and, thank goodness, we've come a long way from the initial command-line tools and will continue to do better.

Go for CLIs

The very same overarching reasons I suggest engineers learn Go are the very same reasons I suggest using Go to build your CLI, but the very history of modern CLIs, which began in the 1960s with the UNIX shell, invented at Bell Labs by Ken Thompson, co-inventor of Golang, holds much weight. The UNIX philosophy, which inspires our modern CLI philosophy and guidelines, is built into the language, and it's clear that many benefits have been born out of this way of thinking:

- **Performance, scalability, and power**

  Golang is quite fast in its compilation and execution. For example, Kubernetes, which is written in Go, has 5 million lines of application code that compile within a couple of minutes. Any other language would take 10 minutes to several hours to compile. Go translates its source code to machine code within its own optimized compiler, allowing easier dependency management. Also, because Golang is still a young language, it's built for contemporary hardware requirements.

  Goroutines are lightweight threads that run concurrently. In my experience programming, I have not seen simplicity and multi-threading go hand in hand, but Golang achieves this extremely well. This feature won my heart.

  The performance and scalability of Go are an obvious draw to the cloud computing community. Many CLIs for cloud computing were written in Go, Docker and Kubernetes included. Any application with a growing user base or a high number of traffic requests needs to consider Golang. Companies such as Uber and Comcast have chosen Go for this reason as well.

- **Simplification of development**

  Golang felt easier to learn than any other language I've ever encountered. The language supports a clean, simple, and fast environment, which is impressive, considering the powerful tools that Go has created. Golang also has many tools that allow the speed and accuracy of development, including formatting tools, testing frameworks, a great linter, and tools to perform static analysis.

- **Versatility**

  Go makes cross-compilation so easy. You can build your application for many different OSs and architectures, increasing accessibility to your CLI. Although executing within different environments, if properly configured, it will work similarly. This will ease users' minds. Later on in this book, we will discuss how to create binaries for different platforms.

- **Growing your skillset**

  Golang is among the most popular languages to learn. Given its many apparent benefits, more start-ups and enterprises are choosing Golang and the demand for Golang developers is growing.

- **Community**

  If you choose Go, you'll be a part of a community of enthusiastic open source developers, willing to partake in the evolutionary journey of a young programming language.

For beginners who are building a CLI in Go for the first time, the next chapters ahead show how clear it is. Golang is an excellent choice for building a CLI, and when there's a need for scalability, performance, and cross compilation, the choice will play in your favor.

Summary

In this chapter, you learned about the scientific discoveries and inventions that led to the creation of the CLI and the remnants of the past that still exist within the terminal today.

Besides a detailed breakdown of the CLI into its parts, this chapter also discussed what a CLI is, overall, and what it is used for. The co-creator of UNIX and Golang, Ken Thompson, influences the philosophy around CLIs and programming in general. This philosophy has been influenced by human and computer interaction over the decades. As with anything with a long history, some ancestral baggage has followed. We learned that in the past, CLIs were primarily written for computers, and today are primarily written for humans. A new element to the UNIX philosophy had to be added: empathy as a cultural norm.

This chapter dug into the guidelines that ultimately embody the UNIX philosophy and why Golang is the best language to implement such a design.

In *Chapter 2, Structuring Go Code for CLI Applications*, we will discuss the first step of project creation: **folder structure**.

Questions

1. What is the TTY on a Linux machine and what is the history behind this command?

2. Which of the core elements of UNIX's philosophy do you resonate with the most? Why?

3. Who are the co-creators of Golang and what is their relationship to the UNIX OS?

4. Can you still visit BBSs today and play some of the old text-based games of the past?

5. Which of the CLI guidelines feels second nature to you and which guidelines would require extra effort?

Answers

1. TTY is a command in UNIX and Linux to display the name of the terminal connected to standard input. TTY is derived from the word teletypewriter, which was the default form of interacting with large mini and mainframe computers.

2. This answer is subjective. However, I like the element of building a prototype first and optimizing second. I prefer the process of building a simple proof of concept that can be used as inspiration. Optimization and refinement can always come later.

3. Golang was created by Robert Griesemer, Rob Pike, and Ken Thompson. Ken Thompson also created the UNIX OS and Rob Pike was a member of the UNIX team.

4. You can still visit BBSs today by downloading a telnet client, for example, CGTerm, and connect to over 1,000 different BBSs still running today. View the list at `https://www.telnetbbsguide.com/`.

5. This answer is subjective. However, I find it second nature to put effort into building constructive help text. Conversely, I think it takes extra effort to write out complete and up-to-date documentation.

Further reading

- PhiloComp.net (`https://philocomp.net/`) is an educational website for both computer scientists and philosophers to learn the links between the disciplines, expanding their view of the other and of themselves

- Command Line Interface Guidelines (`https://clig.dev`) is an excellent resource with plenty of examples of creating CLI applications that adhere strongly to the UNIX philosophy

2

Structuring Go Code for CLI Applications

Programming is like any other creative process. It all begins with a blank slate. Unfortunately, when faced with a blank slate and minimal experience with programming applications from scratch, doubt can kick in – without knowing how to start, you may feel that it's not possible at all.

This chapter is a guide on the first steps of creating a new application, beginning with some of the most popular ways to structure code, describing each, and weighing up their pros and cons. The concept of domain-driven design is discussed, as this can also influence the resulting structure of an application.

An example of an *audio metadata CLI application* gives us an idea of what some real-world use cases or requirements could look like. Learning how to define an application's use cases and requirements is a tedious but necessary step to ensuring a successful project that also meets the needs of all parties involved.

By the end of this chapter, you will have learned all the skills needed to build your application based on your specific use cases and requirements.

This chapter will cover the following topics:

- Commonly used program layouts for robust applications
- Determining use cases and requirements
- Structuring an audio metadata CLI application

Technical requirements

You can find the program layout examples on GitHub at `https://github.com/PacktPublishing/Building-Modern-CLI-Applications-in-Go/tree/main/Chapter02/Chapter-2`.

Commonly used program layouts for robust applications

Along your programming journey, you may come across many different structures for applications. There is no standard programming layout for Go. Given all this freedom, however, the choice of the structure must be carefully made because it will dictate whether we understand and know how to maintain our application. The proper structure for the application will ideally also be simple, easy to test, and directly reflect the business design and how the code works.

When choosing a structure for your Go application, use your best judgment. Do not choose arbitrarily. Listen to the advice in context and learn to justify your choices. There's no reason to choose a structure early, as your code will evolve over time and some structures work better for small applications while others are better for medium to large applications.

Program layouts

Let's dig into some common and emerging structural patterns that have been developed for the Go language so far. Understanding each option will help you choose the best design structure for your next application.

Flat structure

This is the simplest structure to start with and is the most common when you are starting with an application, only have a small number of files, and are still learning about the requirements. It's much easier to evolve a flat structure into a modular structure, so it's best to keep it simple at the start and partition it out later as the project grows.

Let us now see some advantages and disadvantages of this structure:

- **Pros**:

  - It's great for small applications and libraries

  - There are no circular dependencies

  - It's easy to refactor into a modular structure

- **Cons**:

  - This can be complex and disorganized as the project grows

  - Everything can be accessed and modified by everything else

- **Example**

- As the name implies, all the files reside in the root directory in a flat structure. There is no hierarchy or organization and this works well when there is a small number of files:

Figure 2.1 – Example of a flat code structure

As your project grows, there are several different ways to group your code to keep it organized, each with its advantages and disadvantages.

Grouping code by function

Code is separated by its similar functionality. In a *Go REST API* project, as an example, Go files are commonly grouped by handlers and models.

Let us now see some advantages and disadvantages of this structure:

- **Pros**:

    - It's easy to refactor your code into other modular structures

    - It's easy to organize

    - It discourages a global state

- **Cons**:

    - Shared variables or functionality may not have a clear place to live

    - It can be unclear where initialization occurs

To mitigate any confusion that can occur, it's best to follow Go best practices. If you choose the **group-by-function** structure, use the `main.go` file to initialize the application from the project root. This structure, as implied by the name, separates code based on its function. The following figure is an example of groups by function and the types of code that would fall into these different categories:

Group by Function

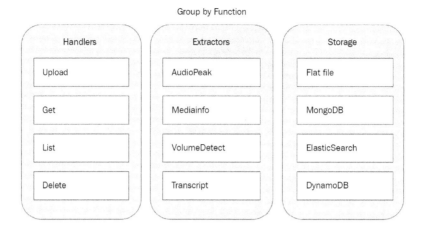

Figure 2.2 – Example of grouping by functionality

- **Example**: The following is an example of folder organization that follows the group-by-function structure. Similar to the example grouping, folders associated with handlers contain code for each type of handler, folders associated with extractors contain code for each particular extraction type, and storage is also organized by type:

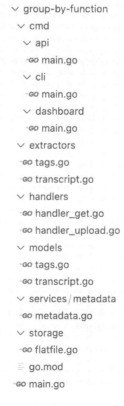

Figure 2.3 – Example of a group-by-function structure

Grouping by module

Unfortunately, the title of this style of architecture is a bit redundant. To clarify, grouping by module means creating individual packages that each serve a function and contain everything necessary to accomplish these functions within them:

- **Pros**:

  - It's easier to maintain

  - There is faster development

  - There is low coupling and high cohesion

- **Cons**:

  - It's complex and harder to understand

  - It must have strict rules to remain well organized

  - It may cause stuttering in package method names

  - It can be unclear how to organize aggregated functionality

  - Circular dependencies may occur

The following is a visual representation of how packages can be grouped by module. In the following example, the code is grouped depending on the implementation of the extractor interface:

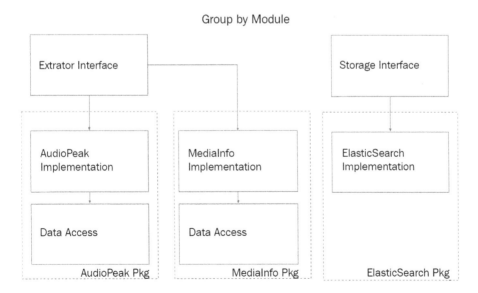

Figure 2.4 – Visual representation of grouping by module

- **Example**

- The following is an example of an organizational structure in which code is grouped into specific module folders. In the following example, the code to extract, store, and define the type, tags, transcript, and other metadata is stored within a single defined folder:

Figure 2.5 – Example of a group-by-module structure

Grouping by context

This type of structure is typically driven by the domain or the specific subject for which the project is being developed. The common domain language used in communication between developers and domain experts is often referred to as ubiquitous language. It helps developers to understand the business and helps domain experts to understand the technical impact of changes.

Hexagonal architecture, also called **ports** and **adapters**, is a popular domain-driven design architecture that conceptually divides the functional areas of an application across multiple layers. The boundaries between these layers are interfaces, also called ports, which define how they communicate with each other, and the adapters exist between the layers. In this layered architecture, the outer layers can only talk to the inner layers, not the other way around:

- **Pros**:

  - There is increased communication between members of the business team and developers

  - It's flexible as business requirements change

  - It's easy to maintain

- **Cons**:

  - It requires domain expertise and for developers to understand the business first before implementation

  - It's costly since it requires longer initial development times

  - It's not suited to short-term projects

The following provides a typical visual representation of a hexagonal structure. The arrows point inward toward entities to distinguish that the outer layers have access to the inner layers, but not the other way around:

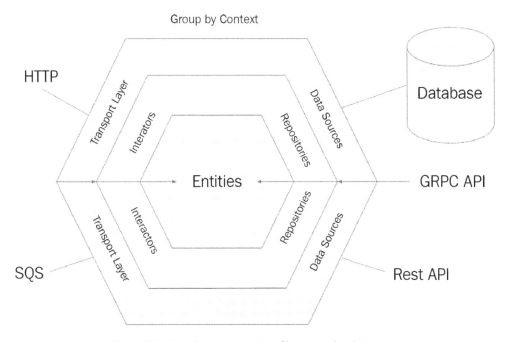

Figure 2.6 – Visual representation of hexagonal architecture

- **Example**: The following is a folder structure organized by context. Services with individual business functions are separated into their respective folders:

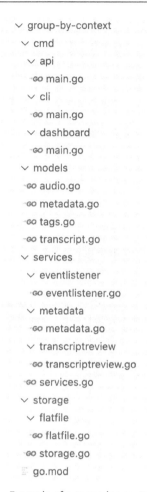

Figure 2.7 – Example of a group-by-context structure

That wraps up the different types of organizational structures for a Go application. There's not necessarily a right or wrong folder structure to use for your application; however, the business structure, the size of the project, and your general preference can play a part in the final decision. This is an important decision to make, so think carefully before moving forward!

Common folders

No matter the chosen structure, there are commonly named folders across existing Go projects. Following this pattern will help to increase understanding for maintainers and future developers of the application:

- **cmd**: The cmd folder is the main entry point for the application. The directory name matches the name of the application.

- **pkg**: The `pkg` folder contains code that may be used by external applications. Although there is debate on the usefulness of this folder, `pkg` is explicit, and being explicit makes understanding crystal clear. I am a proponent of keeping this folder solely due to its clarity.

- **internal**: The `internal` folder contains private code and libraries that cannot be accessed by external applications.

- **vendor**: The `vendor` folder contains the application dependencies. It is created by the `go mod vendor` command. It's usually not committed to a code repository unless you're creating a library; however, some people feel safer having a backup.

- **api**: The `api` folder typically contains the code for an application's *REST API*. It is also a place for Swagger specification, schema, and protocol definition files.

- **web**: The `web` folder contains specific web assets and application components.

- **configs**: The `configs` folder contains configuration files, including any `confd` or `consul-template` files.

- **init**: The `init` folder contains any system initiation (start) and process management (stop/start) scripts with supervisor configurations.

- **scripts**: The `scripts` folder contains scripts to perform various builds, installations, analyses, and operations. Separating these scripts will help to keep the **makefile** small and tidy.

- **build**: The `build` folder contains files for packaging and continuous integration. Any cloud, container, or package configurations and scripts for packaging are usually stored under the `/build/package` folder, and continuous integration files are stored under `build/ci`.

- **deployments** (or **deploy**): The `deployments` folder stores configuration and template files related to system and container orchestration.

- **test**: There are different ways of storing test files. One method is to keep them all together under a `test` folder or to keep the test files right alongside the code files. This is a matter of preference.

> **Note**
>
> No matter what folders are contained within your project structure, use folder names that clearly indicate what is contained. This will help current and future maintainers and developers of the project find what they are looking for. Sometimes, it will be difficult to determine the best name for a package. Avoid overly used terms such as *util*, *common*, or *script*. Format the package name as all lowercase, do not use `snake_case` or `camelCase`, and consider the functional responsibility of the package and find a name that reflects it.

All of the aforementioned common folders and structural patterns described apply to building a CLI application. Depending on whether the CLI is a new feature of an existing application or not, you may be inheriting an existing structure. If there is an existing `cmd` folder, then it's best to define an entry to your CLI there under a folder name that identifies your CLI application. If it is a new

CLI application, start with a flat structure and grow into a modular structure from there. From the aforementioned examples, you can see how a flat structure naturally grows as the application extends to offer more features over time.

Determining use cases and requirements

Before building your CLI application, you'll need to have an idea of the application's purpose and responsibilities. The purpose of the application can be defined as an overarching description, but to start implementing, it's necessary to break down the purpose into use cases and requirements. The goal of use cases and requirements is to drive effective discussion around what an application should do, with the result that everyone has a shared understanding of what is going to be built and continues these discussions as the application evolves.

Use cases

Use cases are a way of documenting the functional requirements of a project. This step, at least for CLIs, is typically handled by an engineer after gathering some high-level requirements from their internal, or external, business customers. It's important to have a clear picture of the application's purpose and document the use cases before any technical implementation because the use cases themselves will not include any implementation-specific language or details about the interface. During any discussion of requirements with customers, topics related to implementation may crop up. Ideally, it's best to steer the conversation back toward use case requirements and handle one thing at a time, staying focused on the right kind of discussion with the right people. The resulting use cases will reflect the goals of the application.

Requirements

Requirements document all the nonfunctional restraints on how the system should perform the use cases. Although a system can still work without meeting these nonfunctional requirements, it may not meet user or consumer expectations as a result. There are some common categories for requirements, each discussed in detail next:

- **Security**

  Security requirements ensure that sensitive information is securely transmitted and that the application adheres to secure coding standards and best practices.

  A few example security requirements include the following:

  - Exclude sensitive data related to sessions and systems from logs

  - Delete unused accounts

  - There are no default passwords in use for the application

- **Capacity**

  Capacity requirements deal with the size of data that must be handled by an application to achieve production goals. Determine the storage requirements of today and how your application will need to scale with increased volume demands. A few examples of capacity requirements include the following:

  - Storage space requirements for logging
  - Several concurrent users can use the application at any given time
  - The limit on the amount of data that can be passed into the application

- **Compatibility**

  Compatibility requirements determine the minimum hardware and operating system requirements for the application to run as expected. Examples include stating the following in your installation requirements:

  - The required architecture
  - All compatible and non-compatible hardware
  - CPU and memory requirements

- **Reliability and availability**

  Reliability and availability requirements define what happens during full or partial failure and set the standard for your application's accessibility. A few examples would include the following:

  - Minimum allowed failures per transaction or time frame
  - Defining accessibility hours for your application

- **Maintainability and manageability**

  Maintainability requirements determine how easily the application can be fixed when a bug is discovered or enhanced when there are new feature requirements. Manageability requirements determine how easily an administrator can manage an application. Examples of maintainability requirements include the following:

  - Bugs must be detected quickly and fixed within an appropriate period
  - The application should maintain compatibility with the latest hardware and operating system versions

- **Scalability**

  Scalability requirements determine the highest workload under which your application can still perform as expected. It is mainly driven by two factors: early software decisions and the infrastructure. Scaling can be horizontal or vertical, where horizontal scaling involves adding more nodes to the system and vertical scaling means adding more memory or faster CPUs to a machine. A couple of examples include the following:

- Several concurrently connected users can use the application with the expected results
- The number of transactions per millisecond is limited

- **Usability**

  Usability requirements determine the quality of the user experience. A few simple examples include the following:

  - The application helps guide users toward the correct usage when they do the wrong thing
  - Help and documentation inform users about new arguments and flags to use
  - During a long operation, users are kept up to date on its progress

- **Performance**

  Performance requirements determine the responsiveness of an application. This includes the following:

  - The minimum required time for users to wait for specific operations to complete
  - Responsiveness to users' actions

- **Environment**

  The environment requirements determine which environments the system will be expected to perform within. A few examples include the following:

  - The required environment variables that must be set
  - Dependencies on third-party software that need to be installed alongside applications

By taking the time to define the use cases and requirements, everyone involved will get a clear picture and have a shared understanding of the purpose and functionality of the application. A shared understanding will lead to a product that benefits in several ways, which we will discuss now.

Disadvantages and benefits of use cases and requirements

Having functional and nonfunctional requirements mapped through use cases and requirements can greatly benefit the outcome of an application.

Here are some disadvantages of determining use cases and requirements:

- It slows down the development process because requirements require time to be properly defined
- Use cases and requirements may change over time

Next, we have some advantages of determining the use cases and requirements:

- It provides the best possible outcome
- Engaging in problem-solving discussions with your team determines potential issues, misuse, or misunderstanding
- It defines the application's goals, future targets, and estimated costs
- You can prioritize each of the requirements

The goal is to gain a level of clarity that helps developers focus on solving the problem with the least amount of ambiguity. Beneficial discussions and collaborative time spent with the team determining the goals of the application are necessary aspects of the process that can be achieved in parallel with defining the use cases and requirements.

Use cases, diagrams, and requirements for a CLI

Let's discuss a theoretical scenario to illustrate how to build use cases and diagrams for a CLI. Suppose there is a large audio company with one particular team that focuses entirely on metadata extraction. This team provides audio metadata to their customers and other internal teams within the same audio company. Currently, they have an API available to anyone within the company's internal network, but an operations team requests a CLI tool. The operations team recognizes the benefit of rapidly building scripts around a CLI application, which could open new opportunities for innovation for the team.

The existing customer-facing API use cases should be similar to the CLI since the implementation and the user interface are not a part of the documentation. Consider the use cases for the metadata team's internal-facing CLI here. For record-keeping, we'll number them and take the first several use cases as examples:

1. Uploading audio
2. Requesting metadata
3. Extract metadata
4. Processing speech to text
5. Requesting speech-to-text transcripts
6. Listing audio metadata in storage
7. Searching audio metadata in storage
8. Deleting audio from storage

For record-keeping, we'll number them and take the first three use cases as examples.

Use case 1 – uploading audio

An authenticated member of the operations team can upload audio by providing a file path. The upload process will automatically save uploads to storage and trigger audio processing to extract the metadata, and the application will respond with a unique ID to use when requesting the metadata.

This use case can be broken down into some common components:

- **The actors** are the end users. This can be defined as a human or another machine process. The primary actor in this example use case is a member of the operations team, but since the team wants to use this CLI for scripting, another machine process is also an actor.

- **Preconditions** are statements that must take place for the use case to occur. In this example, the member must be authenticated before any of the use cases can run successfully. The preconditions in *Figure 2.8* are represented by the solid line with an arrow pointing toward **Verify TLS Certificate**, which confirms through the **Certificate Management Client** that the user is authenticated.

- **Triggers** are events that start another use case. These triggers can be either internal or external. In this example use case, the trigger is external – when a user runs the `upload` command. This use case triggers another use case, *Use case 3, Extract Metadata*, internally to extract metadata from the audio file and save it to storage. This is represented by the **Metadata Extractor** process box.

- When everything happens as intended without exceptions or errors, the **basic flow** is activated. In *Figure 2.8*, the basic flow is a solid line. The user uploads the audio and eventually returns an ID in response. Success!

- The **alternative flow** shows variations of the basic flow, in which errors or exceptions happen. In *Figure 2.8*, the alternative flow is a dotted line. The user uploads the audio, but an error occurs – for example, *the user is invalid* or *the audio file does not exist*.

> Note
>
> The use case diagram for uploading audio is illustrated with the basic flow in a solid line and the alternative flow in a dotted line.

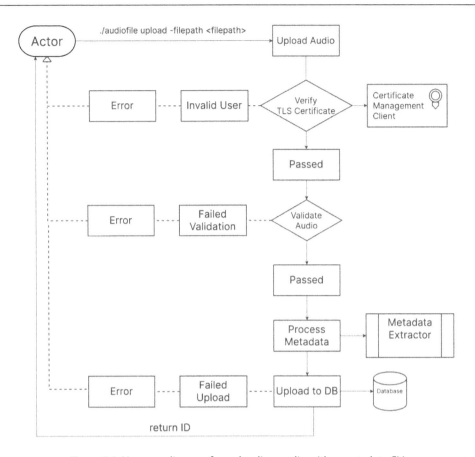

Figure 2.8: Use case diagram for uploading audio with a metadata CLI

Alongside the diagram (*Figure 2.8*), we can write out the use case entirely as follows:

- **Use case 1 – uploading audio:**

  - **Name**: Uploading audio.

  - **Description**: The **Actor** uploads audio by providing a file path. The application returns a unique ID to use for requesting the audio metadata. The actor represents a member of the operations team or another machine process.

  - **Precondition**: The actor must be authenticated. In *Figure 2.8*, this is represented by the solid line from the **Upload Audio** box to the **Verify TLS Certificate** diamond.

  - **Trigger**: The **Actor** triggers the upload command while passing in a valid file path as a flag – in *Figure 2.8*, the arrow pointing from **Actor** to **Upload Audio**.

- **Basic flow:**

  The actor runs the `upload` command in the CLI – in *Figure 2.8*, the arrow pointing from **Actor** to **Upload Audio:**

  I.   Once **Preconditions** have been validated, the audio is vaildated. In *Figure 2.8*, this is represented by the **Validate Audio** box.

  II.   In *Figure 2.8*, the validated audio moves to the **Process Metadata** step, which involves extracting the metadata, this is represented by the arrow pointed to the **Metadata Extractor** process box.

  III.   The validated audio moves to the next step of **Upload Audio**, which saves the audio to the **database** (**DB**), represented by the **Upload to DB** box in *Figure 2.8*.

  IV.   In *Figure 2.8*, the **Return ID** box represents the **ID** being returned from the database, which is later passed down to the **Actor**.

- **Alternative flow:**

  I.   **Error for an unauthenticated user**: An error is returned to the actor when TLS certification fails.

  - **End use case**: In *Figure 2.8*, if the user is invalid, the error is returned, as represented by the dotted line from the **Invalid User** to **Error** box then arrow back to the **Actor**.

  II.   **Error for invalid audio**: An error is returned to the actor when audio fails to pass the validation process.

  - **End use case**: In *Figure 2.8*, if the audio is invalid, an error is returned to the actor, represented by the **Failed Validation** to **Error** box then arrow back to the **Actor**.

  III.   **Error uploading the validated audio to storage**: An error is returned to the actor when audio upload to the database fails.

  - **End use case**: In *Figure 2.8*, the dotted line returned from **Upload to DB** to the **Failed Upload** to **Error** box then arrow back to the **Actor**.

Use case 2 – requesting metadata

An authenticated member of the operations team can retrieve audio metadata by providing an **ID** that was either returned after the upload or found by listing or searching for audio. The `get` command will output the requested audio metadata, with matching ID, in the specified format – either plain text or JSON.

> **Note**
>
> The use case diagram for requesting audio is illustrated with the basic flow in a solid line and the alternative flow in a dotted line.

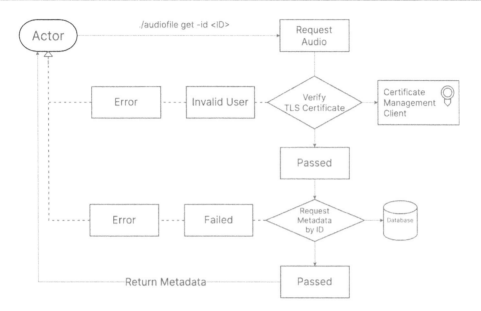

Figure 2.9: Use case diagram for the use case of requesting metadata

With the preceding diagram (*Figure 2.9*) at hand, let's get into the use case as follows:

Use case 2 – requesting metadata:

- **Name**: Requesting metadata.

- **Description**: The **Actor** requests audio metadata by calling the get command and providing an **ID** for the audio. The application will output the requested audio metadata in plaintext or JSON format.

- **Precondition**: The **Actor** must be authenticated. In *Figure 2.9*, this is represented by the solid line from the **Request Audio** box to the **Verify TLS Certificate** diamond.

- **Trigger**: The **Actor** calls the get command while passing in the **ID** as an argument- in *Figure 2.9*, the arrow pointing from **Actor** to the **Request Audio** box.

Note that different formatting levels were used for preceding Use case 1 - make consistent throughout chapter for all use cases?

Basic flow

I.   The actor runs the `get` command in the CLI. In *Figure 2.9*, the basic flow is represented in the solid line and starts with the arrow pointing from **Actor** to the **Request Audio** box.

II.   Once **Preconditions** have been validated, the audio metadata is retrieved by its **ID** from the database. In *Figure 2.9*, this is represented by the solid line connecting **Request Metadata By ID** to **Database**.

III.   The **Database** returns the metadata successfully. In *Figure 2.9*, this is represented by the line connecting **Request Metadata By ID** to the **Passed** box.

IV.   Finally, the formatted metadata is returned to the **Actor**. In *Figure 2.9*, this is represented by the solid line connecting **Passed** to **Actor**.

Alternative flow

I.   **Error for unauthenticated user**: An error is returned to the actor when TLS certification fails.

- **End use case**: In *Figure 2.9*, if the user is invalid, the error is returned, as represented by the dotted line from the **Invalid User** box to the **Error** box then the arrow back to the **Actor**.

II.   **Error for not found**: An error is returned if there is no matching metadata for the **ID**.

- **End use case**: In *Figure 2.9*, the flow is represented by the dotted line from the **Failed** box to the **Error** box and then the arrow back to **Actor**.

Use case 3 – extract metadata

Triggered by **Upload Audio**, metadata, including tags and transcript data, is extracted from the audio file and placed in storage.

> **Note**
>
> The use case diagram for requesting audio is illustrated with the basic flow in a solid line and the alternative flow in a dotted line.

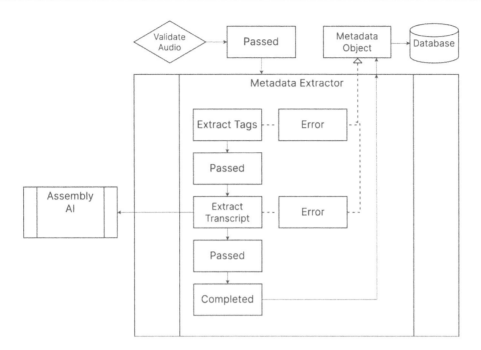

Figure 2.10: Use case diagram for processing metadata

With the preceding diagram (*Figure 2.10*) in mind, let's get into the matching use case:

- **Use case 3 – extract metadata:**

  - **Name**: Extract metadata

  - **Description**: The metadata extraction process consists of extracting specific metadata, including album, artist, year, and speech-to-text transcription, and storing it in a metadata object in storage

  - **Precondition**: Validated audio

  - **Trigger**: Uploading audio

- **Basic flow:**

  I.   Once the **Preconditions** are met, the **Metadata Extractor** process extracts tags and stores the data on the metadata object. In *Figure 2.10*, this is represented by the solid line from the successfully validated audio, with the **Passed** box to the **Metadata Extractor** process box, then from the **Extract Tags** box to the **Passed** box.

  II. Next, the transcript is extracted. In *Figure 2.10*, this is represented by the solid line from the **Passed** box to the **Extract Transcript** box to the next **Passed** box.

III.   The metadata extraction completes and updates the metadata object. In *Figure 2.10*, this step is represented by the solid line from the **Passed** box to **Completed** and the solid line that runs to the **Metadata Object**.

IV.   The metadata object is stored. In *Figure 2.10*, this is represented by the solid line from **Metadata Object** to **Database**.

- **Alternative flow:**

Note that previous two use cases above used different formatting for "End use case" lines - check and make consistent throughout chapter for all use cases

I.   **Error extracting tag data**: In *Figure 2.10*, this is represented by the dotted line from **Extract Tags** to **Error**.

   - **Stores an error on the metadata object**: In *Figure 2.10*, this is represented by the dotted line from **Error** to **Metadata Object**.

   - **End use case**: In *Figure 2.10*, this is represented by the solid line from **Metadata Object** to the **Database**.

II.   **Error extracting the transcript**: An error when extracting transcript metadata occurs. In *Figure 2.10*, this is represented by the dotted line from **Extract Transcript** to **Error**.

   - **Stores an error on the metadata object**: In *Figure 2.10*, this is represented by the dotted line from **Error** to **Metadata Object**.

   - **End use case**: In *Figure 2.10*, this is the solid line from **Metadata Object** to **Database**.

It's not necessary to write out the full documentation for each use case in order to understand the concept. Typically, the functional requirements as described by their use cases are reviewed by the stakeholders, and they are discussed to ensure there is agreement across the board.

Requirements for a metadata CLI

Given our theoretical scenario of an internal team handling all audio metadata, a few nonfunctional requirements may also be requested and defined between the internal team and their customers. The requirements, for example, may include the following:

- The application must run on Linux, macOS, and Windows

- The ID returned from when the audio is uploaded must be returned immediately

- The application must clearly state if the user misuses the application and uploads a file type other than audio

There are many more possible requirements for this metadata CLI application, but it is most important to understand what a requirement is, how to form your own, and how it differs from a use case. Use cases and requirements can be broken down into phases for more granularity, especially for scalability. Applications will grow over time and certain features will be added to match the growing requirements. To reference an earlier CLI guideline, *prototype first and optimize later*, it's best just to get the application working first before optimizing. Depending on the type of issues encountered, whether slow processing, the inability to support a large number of concurrent users, or the inability to handle a certain number of transactions per minute, you will need to resolve them in different ways. For example, you can do load testing when optimizing for concurrent use, or use a memory cache along with a database to optimize the number of transactions handled per minute.

Building a simple prototype for your application can be done in parallel with defining use cases and requirements.

Structuring an audio metadata CLI application

The first step to building a CLI application is creating the folder structure, but if you aren't starting from scratch, determine where the CLI application may be added. Suppose the existing structure for the audio metadata API application was built with a domain-driven architecture. To understand how it may be structured, let's categorize the building blocks of the application.

Bounded context

Bounded context brings a deeper meaning to entities and objects. In the case of our metadata application, consumers utilize the API to search for audio transcription. The operations team would like to search for audio metadata using a CLI. API consumers may be interested in both the metadata and audio transcription but other teams may be more focused on the results of audio transcription. Each team brings a different context to the metadata. However, since tags, album, artist, title, and transcription are all considered metadata, they can be encapsulated within a single entity.

Language

The **language** used to delineate between different contexts is called **ubiquitous language**. Because teams have slightly different meanings for different terms, this language helps to describe the application in terms that are agreed upon by all involved parties.

For the metadata application, the term **metadata** encompasses all the data extracted from audio, including transcription, and **metadata extraction** is the process of extracting technical metadata and transcription from audio. The term **user** refers to any member of an internal team within the larger organization, and the term **audio** to any recorded sound within a specific limit on length.

Entities and value objects

Entities are models of objects defined by the language. Value objects are fields that exist within an entity. For example, the main entities for the metadata CLI are audio and metadata. Metadata is a value object within the audio entity. Also, each extraction type may be its own value object within the Metadata entity. The list of entity and value objects for this audio metadata CLI application includes the following:

- Audio

- Metadata

- Tags

- Transcripts

Aggregation

Aggregation is the merging of two separate entities. Suppose within the metadata team at an audio company users would like to make corrections to transcriptions, which is primarily handled by artificial intelligence. Although the transcription may be 95% accurate, there is a team of reviewers that can make corrections to transcriptions to reach 99-100% accuracy. There would be two microservices within the metadata application, one being metadata extraction and other being transcription review. A new aggregated entity may be required: **TranscriptionReview**.

Service

The term service is generic, so this specifically refers to services within the context of the business domain. In the case of the metadata application, the domain services are the metadata service that extracts metadata from audio and a transcription review service that allows users to add corrections to transcription.

Events

In the context of domain-driven design, events are domain-specific and notify other processes within the same domain of their occurrence. In this particular case, when a user uploads audio, they receive an ID back immediately. However, the metadata extraction process is triggered behind the scenes and rather than continuously polling on the request metadata command or endpoint to retrieve the status of the metadata object, an event can be sent to an event listener service. The CLI could have a command that continuously listens for process completion.

Repository

A repository is a collection of the domain or entity objects. The repository has the responsibility of adding, updating, getting, and deleting objects. It makes aggregation possible. A repository is implemented within the domain layer, so there should be no knowledge of the specific database or storage – within the domain, the repository is only an interface. In the case of this metadata application, the repository can have different implementations – MongoDB, ElasticSearch, or flat file.

Creating the structure

Understanding the components of a domain-driven design, specific to an audio metadata CLI, we can start structuring the folders specific to a metadata CLI. Here is an example layout:

```
/Users/username/go/src/github.com/audiocompany/audiofile
    |--cmd
    |----api
    |----cli
    |--extractors
    |----tags
    |----transcript
    |--internal
    |----interfaces
    |--models
    |--services
    |----metadata
    |--storage
    |--vendor
```

Main folders

Each folder is is follows:

- cmd: The command folder is the main entry point for two different applications that are a part of the audio metadata project: the API and CLI.

- extractors: This folder will hold the packages that will extract metadata from the audio. Although this extractor list will grow, we can start with a few extractor packages: tags and transcript.

- models: This folder will hold all the structs for the domain entities. The main entities to include are audio and metadata. Each of the extractors may also have its own data model and can be stored in this folder.

- services: Three services have been defined in our previous discussion – the metadata (extraction) service, the transcript review service, and an event listener service, which will listen for processing events and output notifications. Existing and new services exist within this folder.

- storage: The interface and individual implementations for storage exist within this folder.

Summary

Throughout this chapter, we have learned how to create a structure for a new application based on the unique requirements of the business domain. We looked at the most popular folder structures for applications and the pros and cons of each, and how to write documentation on use cases and nonfunctional requirements.

While this chapter provided an example layout and the main folders that exist within that example, remember that this is an example of a more developed project. Start simple, always with a flat structure, but start organizing for your future folder structure as you continue to build. Just bear in mind that your code structure will take time. Rome wasn't built in a day.

After covering these topics, we then discussed a hypothetical real-world example of a company with a team focused entirely on audio metadata. We followed this up with some of the potential use cases for a CLI offering, which would be a fast and efficient alternative to the existing API.

Finally, we discussed a folder structure that could satisfy the requirements of the CLI and API audio metadata application. In *Chapter 3*, *Building an Audio Metadata CLI*, we will build out the folder structure with the models, interfaces, and implementations to get the CLI application working. That concludes this chapter on how to structure your Go CLI application! Hopefully, it will help you get started.

Questions

1. If you want to share packages with external applications or users, what common folder would these packages reside in?

2. In ports-and-adapters, or hexagonal, architecture, what are the ports and what are the adapters?

3. For listing audio, in a real-world example, how would you define the actors, preconditions, and triggers of this use case?

Answers

1. The pkg folder contains code that may be used by external applications.

2. Ports are the interfaces and the adapters are the implementations in a hexagonal architecture. Ports allow communication between different layers of the architecture while the adapters provide the actual implementation.

3. The actors are the operations team members or any user of the CLI. A precondition of the use case is that the user must be authenticated first. The use case is triggered by either the API's / list endpoint for the metadata service or running the CLI command for listing audio.

Further reading

- Kat Zein – *How Do you Structure Your Go Apps* from GopherCon 2018 (https://www.youtube.com/watch?v=oL6JBUk6tj0) – an excellent talk about the most common folder structures for Go applications

3

Building an Audio Metadata CLI

Hands-on learning is one of the best ways to learn. So, in this chapter, we will build out a few of our example audio metadata CLI use cases from start to finish. The code is available online and can be explored alongside this chapter or independently. Forking the GitHub repo and playing around with the code, adding in new use cases and tests, are encouraged as these are excellent ways to learn before diving into some of the ways to refine your CLI in the following chapters.

Although this example covered in this chapter is not built on an empty code base – it is built on top of an existing REST API – it's worth noting that the implementation of commands does not necessarily rely on an API. This is only an example and it's encouraged that you use your imagination in this chapter on how commands could be implemented if not relying on an API. This chapter will give you an experimental code base and you'll learn about the following topics:

- Defining the components
- Implementing use cases
- Testing and mocking

Technical requirements

Download the following code to follow along:

```
https://github.com/PacktPublishing/Building-Modern-CLI-Applications-
in-Go/tree/main/Chapter03/audiofile
```

Install the latest version of VS Code with the latest Go tools.

Defining the components

The following is the folder structure for our audio metadata CLI. The main folders in this structure were described in the last chapter. Here, we will go into further detail on what each folder contains, and the files and code that exist within them, in order from top to bottom:

```
|--cmd
|----api
|----cli
|------command
|--extractors
|----tags
|----transcript
|--internal
|----interfaces
|--models
|--services
|----metadata
|--storage
|--vendor
```

cmd/

As previously mentioned in *Chapter 2, Structuring Go Code for CLI Applications* in the *Commonly used program layouts for robust applications* section, the cmd folder is the main entry point for the different applications of the project.

cmd/api/

The main.go file, which in found in the cmd/api/ folder, will start to run the audio metadata API locally on the machine. It takes in a port number as an optional flag, defaulting to 8000, and passes the port number into a Run method within the services method that starts the metadata service:

```
package main

import (
    metadataService "audiofile/services/metadata"
    "flag"
    "fmt"
)
```

```
func main() {
    var port int
    flag.IntVar(&port, "p", 8000, "Port for metadata
      service")
    flag.Parse()
    fmt.Printf("Starting API at http://localhost:%d\n",
      port)
    metadataService.Run(port)
}
```

We make use of the flag package, which implements simple command-line flag parsing. There are different flag types that can be defined, such as String, Bool, and Int. In the preceding example, a -p flag is defined to override the default port of 8000. flag.Parse() is called after all the flags are defined to parse the line into defined flags. There are a few syntactical methods allowed for passing flags to the command using Go's flag package. The value 8080 will be parsed either way:

```
-p=8080
-p 8080   // this works for non-boolean flags only
```

Sometimes, a flag does not require an argument and is enough on its own for the code to know exactly what to do:

```
-p
```

Action can be taken on the flag that's passed in, but the variable will contain the default value, 8000, when defined.

To start the API from the project's root directory, run go run cmd/api/main.go and you will see the following output:

```
audiofile go run cmd/api/main.go
Starting API at http://localhost:8000
```

cmd/cli/

This main.go file, in the cmd/cli/ folder, runs the CLI, and like many other CLIs, this one will utilize the API by making calls to it. Since the API will need to be running for the CLI to work, run the API first in a separate terminal or in the background. The cmd/cli/main.go file contains the following code:

```
package main
```

```
import (
    "audiofile/internal/command"
    "audiofile/internal/interfaces"
    "fmt"
    "net/http"
    "os"
)

func main() {
    client := &http.Client{}
    cmds := []interfaces.Command{
        command.NewGetCommand(client),
        command.NewUploadCommand(client),
        command.NewListCommand(client),
    }
    parser := command.NewParser(cmds)
    if err := parser.Parse(os.Args[1:]); err != nil {
        os.Stderr.WriteString(fmt.Sprintf("error: %v",
          err.Error()))
        os.Exit(1)
    }
}
```

Within the main.go file, the commands are added to a slice of interface Command type. Each command is defined and added:

```
command.NewGetCommand(client),
command.NewUploadCommand(client),
command.NewListCommand(client),
```

Each command takes the client variable, a default http.Client as a parameter to use to make HTTP requests to the audio metadata API endpoints. Passing in the client command allows it to be easily mocked for testing, which we will discuss in the next section.

The commands are then passed into a NewParser method, which creates a pointer to command.Parser:

```
parser := command.NewParser(cmds)
```

This `Parse` function receives all arguments after the application name via the `os.Args[1:]` parameter value. For example, say the command line is called as follows:

```
./audiofile-cli upload -filename recording.m4v
```

Then, the first argument, `os.Args[0]`, returns the following value:

```
audiofile-cli
```

To explain this further, let's look at the `Command` struct and the fields present within it:

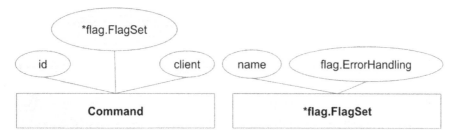

Figure 3.1 – Command struct and flag.FlagSet entities

Let us look at the `GetCommand` struct depicted in the figure:

```
type GetCommand struct {
    fs *flag.FlagSet
    client interfaces.Client
    id string
}
```

Each of the commands has a flag set, which contains a name for the command and error handling, a client, and an ID.

The arguments to a Go program are stored in the `os.Args` slice, which is a collection of strings. The name of the executable being run is stored in the first element of the `os.Args` slice (i.e., `os.Args[0]`), while the arguments passed to the executable are stored in the subsequent elements (`os.Args[1:]`).

When you see the code, `parser.Parse(os.Args[1:])`, it means you're passing the remainder of the command-line arguments to `parse.Parse` function, skipping the first argument (the name of the program). All the arguments on the command line, besides the program's name, will be passed to the function in this case.

That means when we pass in `os.Args[1:]`, we are passing into `parse.Parse` all the arguments after the program name:

```
upload -filename recording.m4v
```

Parse takes `args`, a string list, and returns an `error` type. The function converts command-line parameters into executable commands.

Let's walk through the code alongside the following flow chart:

- It checks for less than 1 args. If so, `help()` returns `nil`.

- `Args[0]` is assigned to subcommand if the slice has at least one item. This shows the user's command.

- The function then cycles over the `Parser` struct's `p.commands` property. It checks each command's name (obtained by executing the `Name()` method) against the `subcommand` variable.

- The function executes the command's `ParseFlags` method with the rest of the `args` slice if a match is found (`args[1:]`). Finally, the function runs the command and returns the result.

- If no match is found, the method returns an unknown subcommand error message using the `fmt.Errorf` function.

Essentially, the code finds and executes a command from command line arguments. Then, the matching command is run.

func (p *Parser) Parse(args []string) error

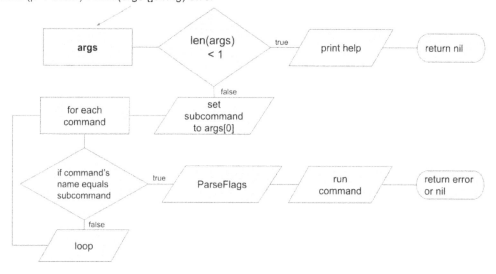

Figure 3.2 – Flow diagram for the Parse method

A command exists for each API endpoint. For example, `UploadCommand` will call the `/upload` endpoint, `ListCommand` will call the `/list` endpoint, and `GetCommand` will call the `/get` endpoint of the REST API.

Within the `Parse` method, the length of `args` is checked. If no arguments are passed, then help is printed and the program returns `nil`:

```
audiofile ./audiofile-cli
usage: ./audiofile-cli <command> [<args>]

These are a few Audiofile commands:
    get       Get metadata for a particular audio file by id
    list      List all metadata
    upload    Upload audio file
```

cmd/cli/command

In the cmd/cli/command folder, there are commands to match each of the audiofile API endpoints. In the next section, we will code the upload, list, and get commands to implement a couple of the use cases described in the previous chapter. Rather than defining the code for one of these commands here, I'll provide a structure used to define a random command that satisfies the Command interface:

```
package command

import (
    "github.com/marianina8/ audiofile/internal/cli"
    "github.com/marianina8/ audiofile/internal/interfaces"
    "flag"
    "fmt"
)

func NewRandomCommand(client interfaces.Client)
    *RandomCommand {
    gc := &RandomCommand{
        fs: flag.NewFlagSet("random",
            flag.ContinueOnError),
        client: client,
    }
    gc.fs.StringVar(&gc.flag, "flag", "", "string flag for
```

```
        random command")
    return gc
}

type RandomCommand struct {
    fs *flag.FlagSet
    flag string
}

func (cmd *RandomCommand) Name() string {
    return cmd.fs.Name()
}

func (cmd *RandomCommand) ParseFlags(flags []string) error {
    return cmd.fs.Parse(flags)
}

func (cmd *RandomCommand) Run() error {
    fmt.Println(rand.Intn(100))
    return nil
}
```

The upload, get, and list commands follow the same structure, but the implementation of the constructor and Run methods differ.

Also, in the cmd/cli/command folder, there is a parser of the struct type with a method to parse the arguments, match them with the commands, and parse any flags found after the subcommand. The NewParser function creates a new instance of the Parser struct. It takes a slice of type [] interfaces.Command as input and returns a pointer to a Parser struct. This initialization method provides an easy way to set up the struct with a set of desired commands. The following is the code inside parser.go:

```
package command
import (
    "github.com/marianina8/audiofile/internal/interfaces"
    "fmt"
)
type Parser struct {
    commands []interfaces.Command
```

```
}
func NewParser(commands []interfaces.Command) *Parser {
    return &Parser{commands: commands}
}
func (p *Parser) Parse(args []string) error {
    if len(args) < 1 {
        help()
        return nil
    }
    subcommand := args[0]
    for _, cmd := range p.commands {
        if cmd.Name() == subcommand {
            cmd.ParseFlags(args[1:])
            return cmd.Run()
        }
    }
    return fmt.Errorf("Unknown subcommand: %s", subcommand)
}
```

The code checks the number of arguments passed to the `Parse` method. If the number of arguments is less than 1, a `help` function from a separate `help.go` file is called to print the help text to guide the user on proper usage:

```
func help() {
    help := `usage: ./audiofile-cli <command> [<flags>]

These are a few Audiofile commands:
    get      Get metadata for a particular audio file by id
    list     List all metadata
    upload   Upload audio file
    `

    fmt.Println(help)
}
```

extractors/

This folder contains implementations for the different extractors of audio metadata. In this case, subfolders exist for the `tags` and `transcript` implementations.

extractors/tags

The `tags` package is implemented within the `extractors/tags` folder. Tags metadata may include title, album, artists, composer, genre, release year, lyrics, and any additional comments. The code is available within the GitHub repository and utilizes the `github.com/dhowden/tag` Go package.

extractors/transcript

The `transcript` package is implemented within the `extractors/transcript` folder. Like the other extraction package, the code can be found in the GitHub repository. However, transcript analysis is handled by AssemblyAI, a third-party API, and requires an API key, which can be set locally to `ASSEMBLY_API_KEY`.

internal/interfaces

The `internal/interfaces` folder holds interfaces utilized by the application. It includes both the `Command` and `Storage` interfaces. Interfaces provide a way for developers to create multiple types that meet the same interface specifications allowing flexibility and modularity in the design of the application. The `storage.go` file defines the storage interface:

```
package interfaces

import (
    "audiofile/models"
)

type Storage interface {
    Upload(bytes []byte, filename string) (string, string,
        error)
    SaveMetadata(audio *models.Audio) error
    List() ([]*models.Audio, error)
    GetByID(id string) (*models.Audio, error)
    Delete(id string, tag string) error
}
```

The preceding interface satisfies all possible use cases. Specific implementations can be defined in the `storage` folder. If you choose to define the storage type within a configuration, you can easily swap out implementations and switch from one storage type to another. In this example, we define flat file storage with an implementation of each method to satisfy the interface.

First utilized in the cmd/cli/main.go file, the Command interface is defined by the following code in internal/interfaces/command.go:

```
type Command interface {
    ParseFlags([]string) error
    Run() error
    Name() string
}
```

Notice how each of the commands in the cmd/cli/command/ folder implements the preceding interface.

models/

The models folder contains a structs shared across the different applications. The first struct defined for the audiofile application is Audio:

```
type Audio struct {
    Id       string
    Path     string
    Metadata Metadata
    Status   string
    Error    []error
}
```

The Id variable contains the unique **identifier(ID)**, of the Audio file. The path the stored local copy of the audio file. The Metadata variable contains the data extracted from the audio file. In the following example, tags and speech-to-text transcript data are being stored:

```
type Metadata struct {
    Tags       Tags      `json:"tags"`
    Transcript string    `json:"transcript"`
}
```

It's not necessary to know the struct for each extraction type. The most important thing is the main entity type, Audio, and its value field, Metadata.

services/metadata

Although multiple services could be implemented in the services folder, we're currently only utilizing one API service, the audio metadata service. The only method that exists in the metadata.go file is the CreateMetadataServer method, which is called in the metadata package, and the Run method, which is called from the `cmd/api/main.go` file. This file also contains the struct for `MetadataService`:

```
type MetadataService struct {
    Server *http.Server
    Storage interfaces.Storage
}
```

`CreateMetadataService` takes an argument, a port of the `int` type, to define the server's port running on localhost. It also takes an argument, `storage`, which is an implementation of the `Storage` interface. The handlers that declare each endpoint of the API server are also defined. This function returns a pointer to `MetadataService`:

```
func CreateMetadataService(port int, storage
    interfaces.Storage) *MetadataService {
    mux := http.NewServeMux()
    metadataService := &MetadataService{
        Server: &http.Server{
            Addr:    fmt.Sprintf(":%v", port),
            Handler: mux,
        },
        Storage: storage,
    }
    mux.HandleFunc("/upload",
      metadataService.uploadHandler)
    mux.HandleFunc("/request",
      metadataService.getByIDHandler)
    mux.HandleFunc("/list", metadataService.listHandler)
    return metadataService
}
```

The Run method, which takes an argument, `port`, defined by the value of the p flag or the default value of `8000`, calls the `CreateMetadataService` method and initiates running the server by calling the `ListenAndServer` method on the server. Any error with starting the API will be returned immediately:

```
func Run(port int) {
    flatfileStorage := storage.FlatFile{}
    service:= CreateMetadataService(port, flatfileStorage)
```

```
    err := service.Server.ListenAndServe()
    if err != nil {
        fmt.Println("error starting api: ", err)
    }
}
```

Implementations of each of the handlers will be discussed in the next section when handling a few use cases.

storage/

In the `storage` folder, there is the `flatfile.go` file, which implements a method of storing metadata locally to a flat file organized via ID on the local disk. The code implementation of this will not be discussed in this book because it goes beyond the scope of focus on the CLI. However, you can view the code in the GitHub repository.

vendor/

The `vendor` directory holds all direct and indirect dependencies.

Implementing use cases

Remember the use cases defined in the previous chapter? Let's try to implement a couple of them:

- UC-01 Upload audio
- UC-02 Request metadata

Uploading audio

In this use case, an authenticated user uploads an audio file by giving the location of the file on their device for the purpose of extracting its metadata. Under the hood, the upload process will save a local copy and run the metadata extraction process on the audio file. A unique ID for the audio file is returned immediately.

Before we begin to implement this use case, let's consider what the command for uploading may look like. Suppose we've settled on the following final command structure:

```
./audiofile-cli upload -filename <filepath>
```

Since /cmd/cli/main.go is already defined, we'll just need to make sure that the upload command exists and satisfies the command interface, with the ParseFlags, Run, and Name methods. In the internal/command folder, we define the upload command in the upload.go file within the command package:

```go
package command

import (
    "github.com/marianina8/audiofile/internal/interfaces"
    "bytes"
    "flag"
    "fmt"
    "io"
    "io"
    "mime/multipart"
    "net/http"
    "os"
    "path/filepath"
)

func NewUploadCommand(client interfaces.Client)
    *UploadCommand {
    gc := &UploadCommand{
        fs:      flag.NewFlagSet("upload",
                    flag.ContinueOnError),
        client: client,
    }

    gc.fs.StringVar(&gc.filename, "filename", "", "full
        path of filename to be uploaded")

    return gc
}

type UploadCommand struct {
    fs       *flag.FlagSet
    client   interfaces.Client
```

```
        filename string
}

func (cmd *UploadCommand) Name() string {
    return cmd.fs.Name()
}

func (cmd *UploadCommand) ParseFlags(flags []string)
  error {
    if len(flags) == 0 {
        fmt.Println("usage: ./audiofile-cli
          upload -filename <filename>")
        return fmt.Errorf("missing flags")
    }
    return cmd.fs.Parse(flags)
}

func (cmd *UploadCommand) Run() error {
    // implementation for upload command
    return nil
}
```

The NewUploadCommand method implements our desired command structure by defining a new flag set for the upload command:

```
flag.NewFlagSet("upload", flag.ContinueOnError)
```

This method call passes the string, upload, into the method's name parameter and flag. ContinueOnError in the flag.ErrorHandling parameter defines how the application should react if an error occurs when parsing the flag. The different, and mostly self-explanatory, options for handling errors upon parsing include the following:

- flag.ContinueOnError
- flag.ExitOnError
- flag.PanicOnError

Now that we've defined and added the upload command, we can test it out. Upon testing, you'll see that the upload command runs without an error but exits immediately with no response. Now, we are ready to implement the Run method of the upload command.

When we first started implementing a CLI for the audiofile application, an API already existed. We discussed how this API starts and runs MetadataServer, which handles requests to a few existing endpoints. For this use case, we are concerned with the http://localhost/upload endpoint.

With this in mind, let's delve deeper into the documentation for the upload endpoint of this REST API so we will know exactly how to construct a curl command.

Uploading audio

In order to upload audio, we'll need to know how to communicate with the API to handle certain tasks. Here are the details required to design a request to handle uploading audio:

- **Method**: POST
- **Endpoint**: http://localhost/upload
- **Header**: Content-Type: multipart/form-data
- **Form data**: Key ("file") Value (bytes) Name (base of filename)

Make sure that the API is running, and then test out the endpoint using curl. Immediately, the ID of the uploaded file is returned:

```
curl --location --request POST 'http://localhost/upload' \
--form 'file=@"recording.mp3"'
8a6dc954-d6df-4fc0-882e-14eb1581d968%
```

After successfully testing out the API endpoint, we can write the Go code that handles the same functionality as the previous curl command within the Run method of UploadCommand.

The new Run method can now be defined. The method supplies the filename that's been passed into the upload command as a flag parameter and saves the bytes of that file to a multipart form POST request to the http://localhost/upload endpoint:

```
func (cmd *UploadCommand) Run() error {
    if cmd.filename == "" {
        return fmt.Errorf("missing filename")
    }
    fmt.Println("Uploading", cmd.filename, "...")
    url := "http://localhost/upload"
    method := "POST"
    payload := &bytes.Buffer{}
    multipartWriter := multipart.NewWriter(payload)
    file, err := os.Open(cmd.filename)
```

```go
if err != nil {
    return err
}
defer file.Close()

partWriter, err := multipartWriter
  .CreateFormFile("file", filepath.Base(cmd.filename))
if err != nil {
    return err
}

_, err = io.Copy(partWriter, file)
if err != nil {
    return err
}

err = multipartWriter.Close()
if err != nil {
    return err
}

client := cmd.client
req, err := http.NewRequest(method, url, payload)
if err != nil {
    return err
}

req.Header.Set("Content-Type",
  multipartWriter.FormDataContentType())
res, err := client.Do(req)
if err != nil {
    return err
}
defer res.Body.Close()

body, err := io.ReadAll(res.Body)
```

```
    if err != nil {
        return err
    }

    fmt.Println("Audiofile ID: ", string(body))
    return err
}
```

The first CLI command, `upload`, has been implemented! Let's implement another use case, requesting metadata by ID.

Requesting metadata

In the requesting metadata use case, an authenticated user requests audio metadata by the audio file's ID. Under the hood, the request metadata process will, within the flat file storage implementation, search for the `metadata.json` file corresponding with the audio file and decode its contents into the `Audio` struct.

Before implementing the request metadata use case, let's consider what the command for requesting metadata will look like. The final command structure will look like this:

./audiofile-cli get -id <ID>

For simplification, `get` is the command to request metadata. Let's define the new `get` command, and in `/cmd/cli/main.go`, confirm that it is present in the list of commands to recognize when the application is run. The structure for defining the `get` command is similar to that of the first command, `upload`:

```
package command

import (
    "github.com/marianina8/audiofile/internal/interfaces"
    "bytes"
    "flag"
    "fmt"
    "io"
    "net/http"
    "net/url"
)
```

```go
func NewGetCommand(client interfaces.Client) *GetCommand {
    gc := &GetCommand{
        fs:      flag.NewFlagSet("get",
                    flag.ContinueOnError),
        client: client,
    }

    gc.fs.StringVar(&gc.id, "id", "", "id of audiofile
      requested")

    return gc
}

type GetCommand struct {
    fs      *flag.FlagSet
    client interfaces.Client
    id      string
}

func (cmd *GetCommand) Name() string {
    return cmd.fs.Name()
}

func (cmd *GetCommand) ParseFlags(flags []string) error {
    if len(flags) == 0 {
        fmt.Println("usage: ./audiofile-cli get -id <id>")
        return fmt.Errorf("missing flags")
    }
    return cmd.fs.Parse(flags)
}

func (cmd *GetCommand) Run() error {
    // implement get command
    return nil
}
```

The `NewGetCommand` method implements our desired command structure by defining a new flag set for the `get` command, `flag.NewFlagSet("get", flag.ContinueOnError)`. This method receives the string, `get`, in the method's name parameter and `flag.ContinueOnError` in the `flag.ErrorHandling` parameter.

Let's delve deeper into the documentation for the get endpoint of this REST API so we will know exactly how to construct a curl command.

Requesting metadata

In order to request audio metadata, we'll need to know how to communicate with the API to handle this task. Here are the details required to design a request for audio metadata:

- **Method**: GET
- **Endpoint**: `http://localhost/get`
- **Query parameter**: `id` – ID of audio file

Make sure that the API is running, and then test out the `get` endpoint using `curl`. Immediately, the metadata of the requested audio file is returned in JSON format. This data could be returned in different formats, and we could add an additional flag to determine the format of the returned metadata:

```
curl --location --request GET
'http://localhost/request?id=270c3952-0b48-4122-bf2a-
 e4a005303ecb'
{audiofile metadata in JSON format}
```

After confirming that the API endpoint works as expected, we can write the Go code that handles the same functionality as the preceding `curl` command within the Run method of `GetCommand`. The new Run method can now be defined:

```
func (cmd *GetCommand) Run() error {
    if cmd.id == "" {
        return fmt.Errorf("missing id")
    }
    params := "id=" + url.QueryEscape(cmd.id)
    path := fmt.Sprintf("http://localhost/request?%s",
      params)
    payload := &bytes.Buffer{}
    method := "GET"
    client := cmd.client
    req, err := http.NewRequest(method, path, payload)
```

```go
    if err != nil {
        return err
    }
    resp, err := client.Do(req)
    if err != nil {
        return err
    }
    defer resp.Body.Close()

    b, err := io.ReadAll(resp.Body)
    if err != nil {
        fmt.Println("error reading response: ",
            err.Error())
        return err
    }
    fmt.Println(string(b))
    return nil
}
```

Now that the request metadata use case has been implemented, let's compile the code and test out the first couple of CLI commands: upload, for uploading and processing audio metadata, and get, for requesting metadata by audiofile ID.

Giving the CLI a more specific name, audiofile-cli, let's generate the build by running the following command:

```
go build -o audiofile-cli cmd/cli/main.go
```

Testing a CLI

Now that we have successfully built the CLI application, we can do some testing to make sure that it's working. We can test out the commands we've created and then write out proper tests to make sure any future changes don't break the current functionality.

Manual testing

To upload an audio file, we'll run the following command:

```
./audiofile-cli upload -filename audio/beatdoctor.mp3
```

The result is as expected:

```
Uploading audio/beatdoctor.mp3 ...
Audiofile ID:   8a6a8942-161e-4b10-bf59-9d21785c9bd9
```

Now that we have the audiofile ID, we can immediately get the metadata, which will change as the metadata updates after each extraction process. The command for requesting metadata is as follows:

```
./audiofile-cli get -id=8a6a8942-161e-4b10-bf59-
9d21785c9bd9
```

The result is the populated Audio struct in JSON format:

```
{
    "Id": "8a6a8942-161e-4b10-bf59-9d21785c9bd9",
    "Path": "/Users/marian/audiofile/8a6a8942-161e-4b10-
bf59-9d21785c9bd9/beatdoctor.mp3",
    "Metadata": {
       "tags": {
           "title": "Shot In The Dark",
           "album": "Best Bytes Volume 4",
           "artist": "Beat Doctor",
           "album_artist": "Toucan Music (Various
             Artists)",
           "genre": "Electro House",
           "comment": "URL: http://freemusicarchive.org/
           music/Beat_Doctor/Best_Bytes_Volume_4/
           09_beat_doctor_shot_in_the_dark\r\nComments:
           http://freemusicarchive.org/\r\nCurator: Toucan
           Music\r\nCopyright: Attribution-NonCommercial
           3.0 International: http://creativecommons.org/
           licenses/by-nc/3.0/"
       },
       "transcript": "This is Sharon."
    },
    "Status": "Complete",
    "Error": null
}
```

The results are as expected. However, not all audio passed into the CLI will return the same data. This is just an example. Some audio may not have any tags at all and transcription will be skipped if you don't have the `ASSEMBLYAI_API_KEY` environment variable set with an AssemblyAI API key. Ideally, API keys should not be set as environment variables, which can be leaked easily, but this is a temporary option. In *Chapter 4, Popular Frameworks for Building CLIs*, you will learn about Viper, which is a configuration library that pairs perfectly with the Cobra CLI framework.

Testing and mocking

Now, we can start writing some unit tests. In the `main.go` file, there is a root function that parses the arguments passed into the application. Using VS Code and the extension for Go support, you can right-click on a function and see an option for generating unit tests, **Go: Generate Unit Tests For Function**.

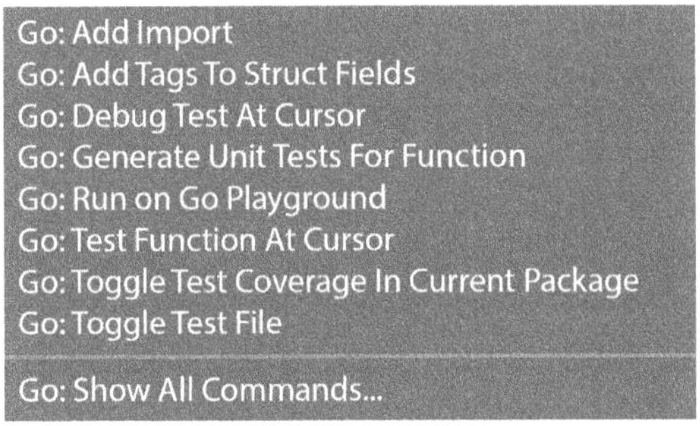

Figure 3.3 –Screenshot of VS Code menu of Go options

Select the `Parse` function in the `commands` package and then click on the option to generate the following table-driven unit tests inside the `parser_test.go` file, we can see the test function for the parsing functionality:

```
func TestParser_Parse(t *testing.T) {
    type fields struct {
        commands []interfaces.Command
    }
    type args struct {
        args []string
    }
    tests := []struct {
        name    string
        fields  fields
        args    args
```

```
        wantErr bool
    }{
        // TODO: Add test cases.
    }
    for _, tt := range tests {
        t.Run(tt.name, func(t *testing.T) {
            p := &Parser{
                commands: tt.fields.commands,
            }
            if err := p.Parse(tt.args.args); (err != nil)
              != tt.wantErr {
                t.Errorf("Parser.Parse() error = %v,
                  wantErr %v", err, tt.wantErr)
            }
        })
    }
}
```

This provides a great template for us to implement some tests given different argument and flag combinations utilized in the method. When running the tests, we don't want the client to call the REST endpoints, so we mock the client and fake responses. We do all this inside the `parser_test.go` file. Since each of the commands takes in a client, we can easily mock the interface. This is done in the file using the following code:

```
type MockClient struct {
    DoFunc func(req *http.Request) (*http.Response, error)
}

func (m *MockClient) Do(req *http.Request) (*http.Response,
    error) {
    if strings.Contains(req.URL.String(), "/upload") {
        return &http.Response{
            StatusCode: 200,
            Body:       io.NopCloser
              (strings.NewReader("123")),
        }, nil
```

```
    }
    if strings.Contains(req.URL.String(), "/request") {
        value, ok := req.URL.Query()["id"]
        if !ok || len(value[0]) < 1 {
            return &http.Response{
StatusCode: 500,
Body: io.NopCloser(strings.NewReader("url param 'id' is
missing")),
}, fmt.Errorf("url param 'id' is missing")
        }
        if value[0] != "123" {
            return &http.Response{
                StatusCode: 500,
                Body:        io.NopCloser
                (strings.NewReader("audiofile id does not
                    exist")),
            }, fmt.Errorf("audiofile id does not exist")
        }
        file, err := os.ReadFile("testdata/audio.json")
        if err != nil {
            return nil, err
        }
        return &http.Response{
            StatusCode: 200,
            Body:        io.NopCloser
                (strings.NewReader(string(file))),
        }, nil
    }
    return nil, nil
}
```

The MockClient interface is satisfied by http.DefaultClient. The Do method is mocked. Within the Do method, we check which endpoint is being called (/upload or /get) and respond with the mock response. In the preceding example, any call to the /upload endpoint responds with an OK status and a string, 123, representing the ID of the audio file. A call to the /get endpoint checks the IDs passed in as a URL parameter. If the ID matches the audiofile ID of 123, then the mocked client will return a successful response with the audio JSON in the body of the response. If there is a request for any ID other than 123, then a status code of 500 is returned with an error message that the ID does not exist.

Now that the mocked client is complete, we fill in success and failure cases for each command, upload and get, within the Parse function's unit tests:

```
func TestParser_Parse(t *testing.T) {
    mockClient := &MockClient{}
    type fields struct {
        commands []interfaces.Command
    }
    type args struct {
        args []string
    }
```

The tests variable contains an array of data that contains the name of the test, the fields or commands available, the string arguments potentially passed into the command-line application, and a wantErr Boolean value that is set depending on whether we expect an error to be returned in the test or not. Let's go over each test:

```
tests := []struct {
        name    string
        fields  fields
        args    args
        wantErr bool
    }{
```

The first test, named upload - failure - does not exist, simulates the following command:

./audiofile-cli upload -filename doesNotExist.mp3

The filename, doesNotExist.mp3, is a file that does not exist in the root folder. Within the Run() method of the upload command, the file is opened. This is where the error occurs and the output is an error message, file does not exist:

```
        {
            name: "upload - failure - does not exist",
            fields: fields{
                commands: []interfaces.Command{
                    NewUploadCommand(mockClient),
                },
            },
            args: args{
```

```
            args: []string{"upload", "-filename",
                "doesNotExist.mp3"},
        },
        wantErr: true, // error = open
            doesNotExist.mp3: no such file or directory
    },
```

The test named upload – success – uploaded checks the successful case of a file being uploaded to storage with an audiofile ID being returned in response. In order to get this test to work, there is a testdata folder in the command package, and within it exists a small audio file to test with, simulating the following command:

`./audiofile-cli upload -filename testdata/exists.mp3`

This file is successfully opened and sent to the /upload endpoint. The mocked client's Do function sees that the request is to the /upload endpoint and sends an OK status along with the audiofile ID of 123 within the body of the response and no error. This matches the wantErr value of false:

```
    {
        name: "upload - success - uploaded",
        fields: fields{
            commands: []interfaces.Command{
                NewUploadCommand(mockClient),
            },
        },
        args: args{
            args: []string{"upload", "-filename", "
                testdata/exists.mp3"},
        },
        wantErr: false,
    },
```

After uploading, we can now *get* the metadata associated with the audiofile. The next test case, get – failure – id does not exist, tests a request for an audiofile ID that does not exist. Instead of passing in 123, that is, that ID of an audiofile that exists, we pass in an ID that does not exist, simulating the following command via the CLI:

`./audiofile-cli get -id 567`

wantErr is set to true and we get the expected error, audiofile id does not exist. The response from the /request endpoint returns the error message in the body of the response.

```
{
    name: "get - failure - id does not exist",
    fields: fields{
        commands: []interfaces.Command{
            NewGetCommand(mockClient),
        },
    },
    args: args{
        args: []string{"get", "-id", "567"},
    },
    wantErr: true, // error = audiofile id does not
      exist
},
```

The test named get - success - requested checks whether the get command was successful in retrieving an ID of an audiofile that exists. The ID passed is "123", and in the mocked client, you can see that when that specific ID is passed into the request, the API endpoint returns a 200 success code with the body of the audiofile metadata.

This is simulated with the following command:

```
./audiofile-cli get -id 123
```

```
{
    name: "get - success - requested",
    fields: fields{
        commands: []interfaces.Command{
            NewGetCommand(mockClient),
        },
    },
    args: args{
        args: []string{"get", "-id", "123"},
    },
    wantErr: false,
},
}
```

The following code loops through the previously described `tests` array to run each test with the arguments passed into the command and checks whether the final `wantErr` value matches the expected error:

```
for _, tt := range tests {
    t.Run(tt.name, func(t *testing.T) {
        p := &Parser{
            commands: tt.fields.commands,
        }
        if err := p.Parse(tt.args.args); (err != nil)
          != tt.wantErr {
            t.Errorf("Parser.Parse() error = %v,
              wantErr %v", err, tt.wantErr)
        }
    })
}
```

To run these tests, from the repository type the following:

```
go test ./cmd/cli/command -v
```

This will execute all the preceding tests and print the following output:

```
--- PASS: TestParser_Parse (0.06s)
    --- PASS: TestParser_Parse/upload_-_failure_-
    _does_not_exist (0.00s)
    --- PASS: TestParser_Parse/upload_-_success_-_uploaded
    (0.06s)
    --- PASS: TestParser_Parse/get_-_failure_-
    _id_does_not_exist (0.00s)
    --- PASS: TestParser_Parse/get_-_success_-_requested
    (0.00s)
PASS
ok          github.com/marianina8/audiofile/cmd/cli/command
(cached)
```

It's important to test success and failure cases for all the commands. Although this was just a starting example; more test cases could be added. For example, in the previous chapter, we discussed the upload use case in more detail. You could test it with large files that exceed the limit, or whether the file passed into the `upload` command is an audio file. In the state that the current implementation is in, a large file would successfully upload. Since this is not what we want, we can modify the `UploadCommand Run` method to check the size of the file before calling the request to the `/upload` endpoint. However, this is just an example and hopefully gives you an idea of how a CLI can be built alongside an existing API.

Summary

Throughout this chapter, we have gone through an example of building an audio metadata CLI. Going through each of the different components that make up this CLI has helped us to determine how a CLI could be structured and how files are structured, whether as part of an existing code base or as a new CLI.

We learned how to implement the first two main use cases of the CLI, uploading audio and getting audio metadata. The details provided on the structure of the commands gave you an idea of how commands could be built out without the use of any additional parsing packages. You also learned how to implement a use case, test your CLI, and mock a client interface.

While this chapter gave you an idea of how to build a CLI, some commands such as nested subcommands and flag combinations can get complicated. In the next chapter, we'll discuss how to use some popular frameworks to help parse complicated commands and improve the CLI development process overall. You'll see how these frameworks can exponentially speed up the development of a new CLI!

Questions

1. What are the benefits of using a storage interface? If you were to use a different storage option, how easy would it be to swap out for the current flat file storage implementation?

2. What's the difference between an argument and a flag? In the following real-world example, what qualifies as an argument or a flag?

    ```
    ./audiofile-cli upload -filename music.mp3
    ```

3. Suppose you'd like to create an additional test for when a user runs the `get` command without passing in any arguments or flags:

    ```
    ./audiofile-cli get
    ```

    What would an additional entry to the `tests` array look like?

Answers

1. Interfaces benefit us when writing modular code that's decoupled and reduces dependency across different parts of the code base. Since we have an interface, it's much easier to swap out the implementation. In the existing code, you'd swap the implementation type in the Run method of the `metadata` package.

2. In this example:

    ```
    ./audiofile-cli upload -filename music.mp3
    ```

    `upload`, `-filename`, and `music.mp3` are all considered arguments. However, flags are specific arguments that are specifically marked by a specific syntax. In this case, `-filename` is a flag.

3. An additional test for when a user runs the `get` command without passing in any arguments or flags would look like this:

    ```
    {
        name: "get - failure - missing required id flag",
        fields: fields{
            commands: cmds,
        },
        args: args{
            args: []string{"get"},
        },
        wantErr: true,
    },
    ```

4

Popular Frameworks for Building CLIs

This chapter will explore the most popular frameworks available to assist you with rapidly developing modern CLI applications. After seeing all the code that it takes to manually create a command and structure a CLI application, you'll see how Cobra allows developers to quickly generate all the scaffolding needed for a CLI application and add new commands easily.

Viper easily integrates with Cobra to configure your applications locally or remotely using multiple formats. The options are extensive, and developers can choose what they feel works best for their project and what they are comfortable with. This chapter will give you an in-depth look at Cobra and Viper through the following topics:

- Cobra – a library for building modern CLI applications

- Viper – easy configuration for CLIs

- Basic calculator CLI using Cobra and Viper

Technical requirements

To easily follow along with the code in this chapter, you will need to do the following:

- Install the Cobra CLI: `https://github.com/spf13/cobra-cli`

- Get the Cobra package: `https://github.com/spf13/cobra`

- Get the Viper package: `https://github.com/spf13/viper`

- Download the following code: `https://github.com/PacktPublishing/Building-Modern-CLI-Applications-in-Go/tree/main/Chapter04`

Cobra – a library for building modern CLI applications

Cobra is a Go library for building powerful and modern CLI applications. It makes defining both simple and complex nested commands easy. The extensive field list for the Cobra `Command` object allows you to access the complete self-documenting help and man pages. Cobra also offers some fun additional benefits, including intelligent shell autocomplete, CLI scaffolding, code generation, and integration with the Viper configuration solution.

The Cobra library provides a much more powerful command structure than one written from scratch. As mentioned, there are many advantages to using the Cobra CLI, so we will dive into a few examples to exhibit its power. Starting a CLI with Cobra from scratch only requires three steps. First, make sure that `cobra-cli` is properly installed. Create a new folder for your project and follow these steps in sequence to set up your new CLI:

1. Change directories into your project folder, `audiofile-cli`:

    ```
    cd audiofile-cli
    ```

2. Create a module and initialize your current directory:

    ```
    go mod init <module path>
    ```

3. Initialize your Cobra CLI:

    ```
    cobra-cli init
    ```

After running just three commands, `ls` shows that the folder structure is already created, and commands are ready to be added. Running the `main.go` file returns the default long description, but once commands are added, the audiofile CLI usage will be displayed with help and examples instead.

If you run `cobra-cli` on its own to see the options available, you'll see there are only four commands, `add`, `completion`, `help`, and `init`. Since we've already used `init` to initialize our project, next, we'll use `add` to create the template code for a new command.

Creating subcommands

The fastest way to add a new command from the Cobra CLI is to run the `cobra-cli` command, `add`. To get more details on this command, we run `cobra-cli add –help`, which shows us the syntax for running the `add` command.

To try to create the example `upload` command from the previous chapter, we would run the following:

```
cobra-cli add upload
```

Let's quickly try calling the code that was generated for the `upload` command:

➡️   `audiofile-cli go run main.go upload`
```
upload called
```

By default, the `upload called` output is returned. Now, let's take a look at the generated code. Within the same file for the command is an `init` function that adds this command to the `root` or `entry` command.

Let's clean this file up and fill in some details for our `upload` command:

```
package cmd

import (
    "github.com/spf13/cobra"
)

// uploadCmd represents the upload command
var uploadCmd = &cobra.Command{
    Use:    "upload [audio|video] [-f|--filename]
      <filename>",
    Short: "upload an audio or video file",
    Long: `This command allows you to upload either an
      audio or video file for metadata extraction.
    To pass in a filename, use the -f or --filename flag
      followed by the path of the file.

    Examples:
    ./audiofile-cli upload audio -f audio/beatdoctor.mp3
    ./audiofile-cli upload video --filename video/
      musicvideo.mp4`,
}

func init() {
    rootCmd.AddCommand(uploadCmd)
}
```

Now, let's create these two new subcommands for the `upload` command to specify either audio or video:

```
    cobra-cli add audio
audio created at /Users/marian/go/src/github.com/
  marianina8/audiofile-cli
    cobra-cli add video
```

```
video created at /Users/marian/go/src/github.com/
  marianina8/audiofile-cli
```

We add audioCmd and videoCmd as subcommands to uploadCmd. The audio command, which contains only the generated code, needs to be modified in order to be recognized as a subcommand. Also, we need to define the filename flag for the audio subcommand. The init function of the audio command will look as follows:

```
func init() {
    audioCmd.Flags().StringP("filename", "f", "", "audio
      file")
    uploadCmd.AddCommand(audioCmd)
}
```

Parsing the filename flag happens within the Run function. However, we want to return an error if the filename flag is missing, so we change the function on audioCmd to return an error and use the RunE method instead:

```
RunE: func(cmd *cobra.Command, args []string) error {
    filename, err := cmd.Flags().GetString("filename")
    if err != nil {
        fmt.Printf("error retrieving filename: %s\n",
        err.Error())
        return err
    }
    if filename == "" {
        return errors.New("missing filename")
    }
    fmt.Println("uploading audio file, ", filename)
    return nil
},
```

Let's try this code out first to see whether we get an error when we don't pass in the subcommand, and when we run the proper example command:

```
cobra-cli add upload
```

We now get an error message relating to the usage of the `upload` command:

```
go run main.go upload
This command allows you to upload either an audio or video
   file for metadata extraction.
     To pass in a filename, use the -f or --filename flag
   followed by the path of the file.


     Examples:
     ./audiofile-cli upload audio -f audio/beatdoctor.mp3
     ./audiofile-cli upload video --filename video/musicvideo.
mp4


Usage:
   audiofile-cli upload [command]


Available Commands:
   audio        sets audio as the upload type
   video        sets video as the upload type
```

Let's correctly run the command with either the shorthand or longhand flag name:

```
cobra-cli add upload audio [-f|--filename]
   audio/beatdoctor.mp3
```

The command then returns the expected output:

```
go run main.go upload audio -f audio/beatdoctor.mp3
uploading audio file,audio/beatdoctor.mp3
```

We've created a subcommand, `audio`, of the `upload` command. Now the implementations for video and audio are called using separate subcommands.

Global, local, and required flags

Cobra allows users to define different types of flags: global and local flags. Let's quickly define each type:

- Global: A global flag is available to the command it is assigned to and every subcommand of that command

- Local: A local flag is only available to the command it is assigned to

Notice that both the video and `audio` subcommands require a flag to parse the `filename` string. It would probably be easier to set this flag as a global flag on `uploadCmd`. Let's remove the flag definition from the `init` function of `audioCmd`:

```go
func init() {
    uploadCmd.AddCommand(audioCmd)
}
```

Instead, let's add it as a global command on `uploadCmd` so that it can also be used by `videoCmd`. The `init` function of `uploadCmd` will now look like this:

```go
var (
    Filename = ""
)

func init() {
    uploadCmd.PersistentFlags().StringVarP(&Filename,
      "filename", "f", "", "file to upload")
    rootCmd.AddCommand(uploadCmd)
}
```

This `PersistentFlags()` method sets a flag as global and persistent for all subcommands. Running the command to `upload` an audio file still works as expected:

```
go run main.go upload audio -f audio/beatdoctor.mp3
uploading audio file,  audio/beatdoctor.mp3
```

In the `audio` subcommand implementation, we check to see whether the filename is set. This is an unnecessary step if we make the file required. Let's change `init` to do that:

```go
func init() {
    uploadCmd.PersistentFlags().StringVarP(&Filename,
      "filename", "f", "", "file to upload")
    uploadCmd.MarkPersistentFlagRequired("filename")
    rootCmd.AddCommand(uploadCmd)
}
```

For local flags, the command would be `MarkFlagRequired("filename")`. Now let's try to run the command without passing in the filename flag:

```
go run main.go upload audio
Error: required flag(s) "filename" not set
```

```
Usage:
  audiofile-cli upload audio [flags]

Flags:
  -h, --help    help for audio

Global Flags:
  -f, --filename string    file to upload


exit status 1
```

An error is thrown by Cobra without having to manually check whether the filename flag is parsed. Because the audio and video commands are subcommands to the `upload` command, they require the newly defined, persistent filename flag. As expected, an error is thrown to remind the user that the filename flag is not set. Another way that your CLI application can help guide users is when they incorrectly type in a command.

Intelligent suggestions

By default, Cobra will provide command suggestions if the user has mistyped a command. An example is when the command is entered:

➔ `go run main.go uload audio`

```
Cobra will automatically respond with some intelligent
  suggestions:
Error: unknown command "uload" for "audiofile-cli"

Did you mean this?
    upload


Run 'audiofile-cli --help' for usage.
exit status 1
```

To disable intelligent suggestions, just add the `rootCmd.DisableSuggestions = true` line to the `init` function for the root command. To change the Levenshtein distance for suggestions, modify the value of `SuggestionsMinimumDistance` on the command. You can also use the `SuggestFor` attribute on a command to explicitly state suggestions, which makes sense for commands that are logical substitutes but aren't close in terms of the Levenshtein distance. Another way to guide first-time users of your CLI is to provide help and man pages for your application. The Cobra framework provides an easy way to automatically generate not only help but also man pages.

Automatically generated help and man pages

As we've already seen, entering a wrong command, or adding the -h or –help flag to the command, will cause the CLI to return the help documentation, automatically generated from the details set within the `cobra.Command` structure. Also, man pages may be generated with the addition of the following import: `"github.com/spf13/cobra/doc"`.

> **Note**
>
> Specifics on how to generate man page documentation will be detailed in *Chapter 9, The Empathic Side of Development*, which includes how to write proper help and documentation.

Powering your CLI

As you can see, there are many benefits to using the Cobra library to power your CLI, giving many features by default. The library also comes with its own CLI for generating scaffolding for a new application and for adding commands, which, with all the options available in the `cobra.Command` struct, gives you everything needed to build a robust and highly customizable CLI.

Compared to writing your CLI from scratch without a framework, you can save hours of your time with many of the built-in advantages: command scaffolding, excellent command, argument, and flag parsing, intelligent suggestions, and autogenerated help text and man pages. You can also pair your Cobra CLI with Viper to configure your application for additional benefits.

Viper – easy configuration for CLIs

Steve Francia, author of Cobra, also created a configuration tool, Viper, to easily integrate with Cobra. For a simple application that you're running locally on your machine, you may not initially need a configuration tool. However, if your application may run within different environments that require different integrations, API keys, or general customizations that are better in a config file versus hardcoded, Viper will help ease the process of configuring your app.

Configuration types

There are many ways Viper allows you to set your application's configuration:

- Reading from configuration files
- With environment variables
- With remote config systems
- With command-line flags
- With a buffer

The configuration formats accepted from these configuration types include JSON, TOML, YAML, HCL, INI, envfile, and Java properties formats. To get a better understanding, let's go over an example of each configuration type.

Config file

Suppose we have different URL and port values to connect to depending on different environments. We could set up a YAML configuration file, `config.yml`, that looks like this and is stored within the main folder of our application:

```
environments:
  test:
    url: 89.45.23.123
    port: 1234
  prod:
    url: 123.23.45.89
    port: 5678
loglevel: 1
keys:
  assemblyai: xxxxxxxxxxxxxxxxxxxxxxxxxxxxxxxx
```

Use the code to read in the configuration and test, printing out the prod environment's URL:

```
viper.SetConfigName("config) // config filename, omit
  extension
viper.AddConfigPath(".")      // optional locations for
  searching for config files
err = viper.ReadInConfig()    // using the previous
  settings above, attempt to find and read in the
    configuration
if err != nil { // Handle errors
    panic(fmt.Errorf("err: %w \n", err))
}
fmt.Println("prod environment url:",
  viper.Get("environments.prod.url"))
```

Running the code confirms that `Println` will return `environments.prod.url` as `123.23.45.89`.

Environment variable

Configuration may also be set via environment variables; just note that Viper's recognition of environment variables is case sensitive. There are a few methods that can be used when working with environment variables.

`SetEnvPrefix` tells Viper that the environment variables used with the `BindEnv` and `AutomaticEnv` methods will be prefixed with a specific unique value. For example, say the test URL is set within an environment variable:

```
viper.SetEnvPrefix("AUDIOFILE")
viper.BindEnv("TEST_URL")
os.Setenv("AUDIOFILE_TEST_URL", "89.45.23.123") //sets the
  environment variable
fmt.Println("test environment url from environment
  variable:", viper.Get("TEST_URL"))
```

As mentioned, the prefix, `AUDIOFILE`, affixes to the start of each environment variable passed into the `BindEnv` or `Get` method. When the preceding code is run, the value printed for the test environment URL from the `AUDIOFILE_TEST_URL` environment variable is `89.45.23.123`, as expected.

Command-line flags

Viper supports configuration via several different types of flags:

- Flags: Flags defined using the standard Go library flag package
- Pflags: Flags defined using Cobra/Viper's `pflag` definition
- Flag interfaces: Custom structures that satisfy a flag interface required by Viper

Let's check out each of these in detail.

Flags

Building on top of the standard Go flag package. The Viper `flags` package extends the functionality of the standard flag package, providing additional features such as environment variable support and the ability to set default values for flags. With Viper flags, you can define flags for string, Boolean, integer, and floating-point types, as well as for arrays of these types.

Some example code may look as follows:

```
viper.SetDefault("host", "localhost")
viper.SetDefault("port", 1234)
```

```
viper.BindEnv("host", "AUDIOFILE_HOST")
viper.BindEnv("port", "AUDIOFILE_PORT")
```

In the preceding example, we set default values for the "host" and "port" flags and then bind them to environment variables using viper.BindEnv. After setting the environment variables, we can access the values of the flags using viper.GetString("host") and viper.GetInt("port").

Pflags

pflag is the flag package specific to Cobra and Viper. The values can be parsed and bound. viper.BindPFFlag, for individual flags, and viper.BindPFFlags, for flag sets, are used to bind the value of the flag when it is accessed rather than defined.

Once the flags are parsed and bound, the values can be accessed anywhere in the code using Viper's Get methods. For retrieving the port, we'd use the following code:

```
port := viper.GetInt("port")
```

Within the init function, you can define a command-line flag set and bind the values once they are accessed. Take the following example:

```
pflag.CommandLine.AddGoFlagSet(flag.CommandLine)
pflag.Int("port", 1234, "port")
pflag.String("url", "12.34.567.123", "url")
plag.Parse()
viper.BindPFlags(pflag.CommandLine)
```

Flag interfaces

Viper also allows custom flags that satisfy the following Go interfaces: FlagValue and FlagValueSet.

The FlagValue interface is as follows:

```
type FlagValue interface {
    HasChanged() bool
    Name() string
    ValueString() string
    ValueType() string
}
```

The second interface that Viper accepts is `FlagValueSet`:

```
type FlagValueSet interface {
    VisitAll(fn func(FlagValue))
}
```

An example of code that satisfies this interface is as follows:

```
type customFlagSet struct {
    flags []customFlag
}

func (set customFlagSet) VisitAll(fn func(FlagValue)) {
    for i, flag := range set.flags {
fmt.Printf("%d: %v\n", i, flag)
        fn(flag)
    }
}
```

Buffer

Finally, Viper allows users to configure their applications with a buffer. Using the same value that exists within the configuration file in the first example, we pass the YAML data in as a raw string into a byte slice:

```
var config = []byte(`
    environments:
    test:
      url: 89.45.23.123
      port: 1234
    prod:
      url: 123.23.45.89
      port: 5678
  loglevel: 1
  keys:
    assemblyai: ad915a598023092382348923904
82304
`)
viper.SetConfigType("yaml")
```

```
viper.ReadConfig(bytes.NewBuffer(config))
viper.Get("environments.test.url") // 89.45.23.123
```

Now that you know the different types or ways of configuring your command-line application – from a file, environment variable, flags, or buffer – let's see how to watch for live changes on these configuration types.

Watching for live config changes

Both remote and local configurations can be watched. After making sure all configuration paths are added, call the WatchConfig method to watch for any live changes and take action by implementing a function to pass into the OnConfigChange method:

```
viper.OnConfigChange(func(event fsnotify.Event) {
    fmt.Println("Config modified:", event)
})
viper.WatchConfig()
```

To watch for changes on a remote config, first, read in the remote config using ReadRemoteConfig(), and on the instance of the Viper configuration, call the WatchRemoteConfig() method. Some sample code is as follows:

```
var remoteConfig = viper.New()
remoteConfig.AddRemoteProvider("consul",
    "http://127.0.0.1:2380", "/config/audiofile-cli.json")
remoteConfig.SetConfigType("json")
err := remoteConfig.ReadRemoteConfig()
if err != nil {
    return err
}
remoteConfig.Unmarshal(&remote_conf)
```

The following is an example of a goroutine that will continuously watch for remote configuration changes:

```
go func(){
    for {
        time.Sleep(time.Second * 1)
        _ := remoteConfig.WatchRemoteConfig()
        remoteConfig.Unmarshal(&remote_conf)
```

```
    }
} ()
```

I think that there's much to benefit from utilizing a configuration library rather than starting from scratch, which again can save you hours and expedite your development process. Besides the different ways you can configure your application, you can also provide remote configuration and watch for any changes live. This further creates a more robust application.

Basic calculator CLI using Cobra and Viper

Let us pull some of the pieces together and create a separate and simple CLI using the Cobra CLI framework and Viper for configuration. A simple idea that we can easily implement is a basic calculator that can add, subtract, multiply, and divide values. The code for this demo exists within the Chapter-4-Demo repository for you to follow along.

The Cobra CLI commands

The commands are created with the following cobra-cli command calls:

```
cobra-cli add add
cobra-cli add subtract
cobra-cli add multiply
cobra-cli add divide
```

Calling these commands successfully generates the code for each command, ready for us to fill in the details. Let us show each command and how they each are similar and different.

The add command

The add command, addCmd, is defined as a pointer to the cobra.Command type. Here, we set the fields for the command:

```
// addCmd represents the add command
var addCmd = &cobra.Command{
    Use: "add number",
    Short: "Add value",
    Run: func(cmd *cobra.Command, args []string) {
        if len(args) > 1 {
            fmt.Println("only accepts a single argument")
            return
        }
```

```
        if len(args) == 0 {
            fmt.Println("command requires input value")
            return
        }
        floatVal, err := strconv.ParseFloat(args[0], 64)
        if err != nil {
            fmt.Printf("unable to parse input[%s]: %v",
              args[0], err)
            return
        }
        value = storage.GetValue()
        value += floatVal
        storage.SetValue(value)
        fmt.Printf("%f\n", value)
    },
}
```

Let us take a quick walk through the Run field, which is a first-class function. Before doing any calculations, we check args. The command only takes one numerical field; any more or less will print a usage statement and return the following:

```
if len(args) > 1 {
    fmt.Println("only accepts a single argument")
    return
}
if len(args) == 0 {
    fmt.Println("command requires input value")
    return
}
```

We take the first and only argument, return it, set it within args[0], and parse it to a flat variable using the following code. If the conversion to a float64 value fails, then the command prints out a message about being unable to parse the input and then returns:

```
floatVal, err := strconv.ParseFloat(args[0], 64)
if err != nil {
    fmt.Printf("unable to parse input[%s]: %v", args[0],
      err)
```

```
      return
  }
```

If the conversion is successful, and no errors are returned from the string conversion, then we have a value set for `floatVal`. In our basic calculator CLI, we are storing the value in a file, which is the simplest way to store it for this example. The `storage` package and how Viper is used in configuration will be discussed after the commands. At an elevated level, we get the current value from storage and apply the operation to `floatVal`, and then save it back into storage:

```
value = storage.GetValue()
value += floatVal
storage.SetValue(value)
```

Last but not least, the value is printed back to the user:

```
fmt.Printf("%f\n", value)
```

That concludes our look at the Run function of the add command. The `Use` field describes the usage, and the `Short` field gives a brief description of the command. This concludes the walk-through of the add command. The subtract, multiply, and divide Run functions on their respective commands are remarkably similar, so I will just point out some differences to note.

The subtract command

The same code is used for `subtractCmd`'s Run function with just a small exception. Instead of adding the value to `floatVal`, we subtract it with the following line:

```
value -= floatVal
```

The multiply command

The same code is used for `multiplyCmd`'s Run function, except we multiply it with the following line:

```
value *= floatVal
```

The divide command

Finally, the same code is used for `divideCmd`'s Run function, except for dividing it by `floatVal`:

```
value /= floatVal
```

The clear command

The clear command resets the stored value to 0. The code for clearCmd is short and simple:

```
// clearCmd represents the clear command
var clearCmd = &cobra.Command{
    Use: "clear",
    Short: "Clear result",
    Run: func(cmd *cobra.Command, args []string) {
        if len(args) > 0 {
            fmt.Println("command does not accept args")
            return
        }
        storage.SetValue(0)
        fmt.Println(0.0)
    },
}
```

We check whether any args are passed in, and if so, we print that the command does not accept any arguments and return. If the command is called ./calculator clear, then the 0 value is stored and then printed back to the user.

The Viper configuration

Let's now discuss a simple way to use Viper configuration. In order to keep track of the value that has operations applied to it, we need to store this value. The easiest way to store data is in a file.

The storage package

Within the repository, there's a file, storage/storage.go, with the following code to set the value:

```
func SetValue(floatVal float64) error {
    f, err := os.OpenFile(viper.GetString("filename"),
        os.O_RDWR|os.O_CREATE|os.O_TRUNC, 0755)
    if err != nil {
        return err
    }
    defer f.Close()
    _, err = f.WriteString(fmt.Sprintf("%f", floatVal))
    if err != nil {
```

```
        return err
    }
    return nil
}
```

This code will write the data to the filename returned from `viper.GetString("filename")`. The code to get the value from the file is defined with the following code:

```
func GetValue() float64 {
    dat, err := os.ReadFile(viper.GetString("filename"))
    if err != nil {
        fmt.Println("unable to read from storage")
        return 0
    }
    floatVal, err := strconv.ParseFloat(string(dat), 64)
    if err != nil {
        return 0
    }
    return floatVal
}
```

Again, the same method is used to get the filename, to read, parse, and then return the data contained.

Initializing the configuration

Inside the `main` function, we call the Viper methods to initialize our configuration right before we execute the command:

```
func main() {
    viper.AddConfigPath(".")
    viper.SetConfigName("config")
    viper.SetConfigType("json")
    err := viper.ReadInConfig()
    if err != nil {
        fmt.Println("error reading in config: ", err)
    }
    cmd.Execute()
}
```

> **Note**
>
> The AddConfigPath method is used to set the path for Viper to search for the configuration file. The SetConfigName method allows you to set the name of the configuration file, without the extension. The actual configuration file is config.json, but we pass in config. Finally, the ReadInConfig method reads in the configuration to make it available throughout the application.

The configuration file

Finally, the configuration file, config.json, stores the value for the filename:

```json
{
    "filename": "storage/result"
}
```

This file location can work for a UNIX- or Linux-based system. Change this to suit your platform and try the demo out for yourself!

Running the basic calculator

To quickly build the basic calculator on UNIX or Linux, run go build -o calculator main. go. On Windows, run go build -o calculator.exe main.go.

I ran this application on my UNIX-based terminal and got the following output:

```
% ./calculator clear
0
% ./calculator add 123456789
123456789.000000
% ./calculator add 987654321
1111111110.000000
% ./calculator add 1
1111111111.000000
% ./calculator multiply 8
8888888888.000000
% ./calculator divide 222222222
40.000000
% ./calculator subtract 40
0.000000
```

Hopefully, this simple demo has provided you with a good understanding of how you can use the Cobra CLI to help speed up development and Viper for a simple way to configure your application.

Summary

This chapter took you through the most popular library for building modern CLIs – Cobra – and its partner library for configuration – Viper. The Cobra package was explained in detail and the CLI's usage with examples was also described. We went through examples to take you through generating your initial application code with the Cobra CLI, adding new commands and modifying the scaffolding, to autogenerate useful help and man pages. Viper, as a configuration tool that fits perfectly alongside Cobra, was also described, along with many of its options, in detail.

In the next chapter, we'll discuss how to handle input to a CLI – whether it's text in the form of commands, arguments, or flags, or the control characters that allow you to quit out of a terminal dashboard. We'll also discuss different ways this input is processed and how to output results back to the user.

Questions

1. If you want to have a flag that is accessible to a command and all its subcommands, what kind of flag would be defined and how?

2. What formatting options does Viper accept for configuration?

Answers

1. A global flag using the `PersistentFlag()` method when defining a flag on a command.

2. JSON, TOML, YAML, HCL, INI, envfile, and Java properties formats.

Further reading

- *Cobra – A Framework for Modern CLI Apps in Go* (`https://cobra.dev/`) provides extensive documentation for Cobra with examples utilizing Cobra and links to the Viper documentation

Part 2:
The Ins and Outs of a CLI

This part focuses on the anatomy of a command-line application and the different types of inputs it can receive, such as subcommands, arguments, and flags, as well as other inputs such as stdin, signals, and control characters. It also covers various methods for processing data and how to return the result, including handling errors and timeouts when interacting with external commands or API services. The chapter also highlights Go's cross-platform capabilities using packages such as os, time, path, and runtime.

This part has the following chapters:

- *Chapter 5, Defining the Command-Line Process*
- *Chapter 6, Calling External Processes, Handling Errors and Timeouts*
- *Chapter 7, Developing for Different Platforms*

5

Defining the Command-Line Process

At the core of a command-line application is its ability to process user input and return a result that either a user can easily comprehend or that another process can read as standard input. In *Chapter 1, Understanding CLI Standards*, we discussed the anatomy of a command-line application, but this chapter will go into detail on each aspect of its anatomy, breaking down the different types of input: subcommands, arguments, and flags. Additionally, other inputs will be discussed: `stdin`, signals, and control characters.

Just as there are many types of input that a command-line application can receive, there are many types of methods for processing data. This chapter won't leave you hanging – examples of processing for each input type will follow.

Finally, it's just as important to understand how to return the result, either data if successful or an error on failure, in a way that both humans and computers can easily interpret.

This chapter will cover how to output the data for each end user and the best practices for CLI success. We will cover the following topics:

- Receiving the input and user interaction
- Processing data
- Returning the resulting output and defining best practices

Technical requirements

To easily follow along with the code in this chapter, you will need to do the following:

Download the following code: `https://github.com/PacktPublishing/Building-Modern-CLI-Applications-In-Go/tree/main/Chapter05/application`

Receiving the input and user interaction

The primary methods for receiving input via a command-line application are through its subcommands, arguments, and options, also known as **flags**. However, additional input can come in the form of stdin, signals, and control characters. In this section, we'll break down each different input type and when and how to interact with the user.

Defining subcommands, arguments, and flags

Before we start characterizing the main types of input, let's reiterate the structural pattern that explains the generalized location for each input type in terms of its predictability and familiarity. There's an excellent description of the pattern within the **Cobra Framework documentation**. This is one of the best explanations because it compares the structure to natural language and, just like speaking and writing, the syntax needs to be properly interpreted:

```
APPNAME NOUN VERB -ADJECTIVE
```

> **Note**
>
> The **argument** is the noun and the **command or subcommand(s)** is the verb. Like any modifier, the **flag** is an adjective and adds description.

> **Note**
>
> Most other programming languages suggest using two dashes instead of one. Go is unique in the fact that the single dash and double dash are equivalent to the internal flag package. It is important to note, however, that the Cobra CLI flag does differentiate between single and double dashes, where a single dash is for a short-form flag, and the double dash is for a long-form flag.

In the preceding example, the command and argument, or NOUN VERB, can also be ordered as VERB NOUN. However, NOUN VERB is more commonly used. Use what makes sense to you:

```
APPNAME ARGUMENT <COMMAND | SUBCOMMANDS> --FLAG
```

You might run up against limitations depending on your command-line parser. However, if possible, make arguments, flags, and subcommands order-independent. Now, let's define each in more detail next and use **Cobra** to create a command that utilizes each input type.

Commands and subcommands

At a very basic level, a command is a specific instruction given to a command-line application. In the pattern we just looked at, these are verbs. Think of the way we naturally speak. If we were to talk to a dog, we'd give it commands such as *"roll over," "speak,"* or *"stay."* Since you define the application, you can choose the verbs to define instructions. However, the most important thing to remember when choosing a command (and subcommand) is for names to be clear and consistent.

Ambiguity can cause a lot of stress for a new user. Suppose you have two commands: `yarn update` and `yarn upgrade`. For a developer who is using `yarn` for the first time, do you think it's clear how these commands are different? Clarity is paramount. Not only does it make your application easier to use but it also puts your developer at ease.

As you gain a broad view of your application, you can intuitively determine more clear and more concise language when defining your commands. If your application feels a bit complex, you can utilize subcommands for simplification, and whenever possible, use familiar words for both commands and subcommands.

Let's use the **Docker** application as an example of how subcommands are clearly defined. Docker has a list of management commands such as the following:

- `container` to manage containers
- `image` to manage images

You'll notice that when you run `docker container` or `docker image`, the usage is printed out, along with a list of subcommands, and you'll also notice that there are several subcommands used across these two commands. They remain consistent.

Users of Docker know that the action (`ls`, `rm`, or `inspect`, for example) is related to the subject (image or `container`). The command follows the expected pattern of `"APPNAME ARGUMENT COMMAND"` – `docker image ls` and `docker container ls` too. Notice that `docker` also uses familiar Unix commands – `ls` and `rm`. Always use a familiar command where you can.

Using the Cobra CLI, let's make two commands, with one as a subcommand of the other. Here's the first command we'll add:

```
cobra-cli add command
command created at /Users/marian/go/src/github.com/
   marianina8/application
```

Then, add the subcommand:

```
cobra-cli add subcommand
subcommand created at /Users/marian/go/src/github.com/
   marianina8/application
```

Then, create it as a subcommand by modifying the default line to run AddCommand on commandCmd:

```
func init() {
    commandCmd.AddCommand(subcommandCmd)
}
```

The Cobra CLI makes it incredibly easy not only to create commands but also subcommands as well. Now, when the command is called with the subcommand, we get confirmation that the subcommand is called:

```
./application command subcommand
subcommand called
```

Now, let us understand arguments.

Arguments

Arguments are nouns – things – that are acted upon by the command. They are positional to the command and usually come before the command. The order is not strict; just be consistent with the order throughout your application. However, the very first argument is the application name.

Multiple arguments are okay for actions against multiple files, or multiple strings of input. Take, for example, the rm command and removing multiple files. For example, rm arg1.txt arg2.txt arg3.txt would act on (by removing) the multiple files listed after the command. Allow globbing where it makes sense. If a user wants to remove all the text files in the current directory, then an example of rm *.txt would also be expected to work. Now, consider the mv command, which requires two arguments for the source and target files. An example of mv old.txt new.txt will move old.txt, the source, to the target, new.txt. Globs may also be used with this command.

> **Note**
> Having multiple arguments for *different* things might mean rethinking the way that you're structuring your command. It could also mean that you could be utilizing flags here instead.

Again, familiarity plays in your favor. Use the standard name if there is one and your users will thank you. Here are examples of some common arguments: history, tag, volume, log, and service.

Let's modify the subcommand's generated Run field to identify and print out its arguments:

```
Run: func(cmd *cobra.Command, args []string) {
    if len(args) == 0 {
        fmt.Println("subcommand called")
    } else {
        fmt.Println("subcommand called with arguments: ",
            args)
    }
},
```

Now, when we run the same subcommand with arguments, the following output is printed out:

```
./application command subcommand argument1 argument2
subcommand called with arguments:  [argument1 argument2]
```

Interestingly, flags can provide more clarity over arguments. In general, it does require more typing, but flags can make it more clear what's going on. Another additional benefit is if you decide to make changes to how you receive input, it's much easier to add or remove a flag than it is to modify an existing command, which can break things.

Flags

Flags are adjectives that add a description to an action or command. They are named parameters and can be denoted in different ways, with or without a user-specified value:

- A **hyphen with a single-letter name** (-h)

- A **double-hyphen with a multiple-letter name** (--help)

- A **double-hyphen with a multiple-letter name and a user-specified value** (--file audio. txt, or --file=audio.txt)

It's important to have full-length versions of all flags – single letters are only useful for commonly used flags. If you use single letters for all available flags, there may be more than one flag that starts with that same letter, and that single letter would make sense intuitively for more than one flag. This can add confusion, so it's best not to clutter the list of single-letter flags.

Single-letter flags may also be concatenated together. For example, take the ls command. You can run ls -l -h -F or ls -lhF and the result is the same. Obviously, this depends on the command-line parser used, but because CLI applications typically allow you to concatenate single-letter flags, it's a good idea to allow this as well.

Finally, the flag order is typically not strict, so whether a user runs ls –lhF, ls –hFl, or ls – Flh, the result is the same.

As an example, we can add a couple of flags to the root command, one local and another persistent, meaning that it's available to the command and all subcommands. In commandCmd, within the init() function, the following lines do just that:

```
commandCmd.Flags().String("localFlag", "", "a local string
   flag")
commandCmd.PersistentFlags().Bool("persistentFlag", false,
   "a persistent boolean flag")
```

In commandCmd's Run field, we add these lines:

```
localFlag, _ := cmd.Flags().GetString("localFlag")
if localFlag != "" {
    fmt.Printf("localFlag is set to %s\n", localFlag)
}
```

In `subcommandCmd`'s Run field, we also add the following lines:

```
persistentFlag, _ := cmd.Flags().GetBool("persistentFlag")
fmt.Printf("persistentFlag is set to %v\n", persistentFlag)
```

Now, when we compile the code and run it again, we can test out both flags. Notice that there are multiple ways of passing in flags and in both cases, the results are the same:

```
  ./application command --localFlag="123"
command called
localFlag is set to 123
  ./application command --localFlag "123"
command called
localFlag is set to 123
```

The persistent flag, although defined at the `commandCmd` level, is available within `subcommandCmd`, and when the flag is missing, the default value is used:

```
  ./application command subcommand
subcommand called
persistentFlag is set to false
  ./application command subcommand --persistentFlag
subcommand called
persistentFlag is set to true
```

Now, we've covered the most common methods of receiving input to your CLI: commands, arguments, and flags. The next methods of input include piping, signal and control characters, and direct user interaction. Let's dive into these now.

Piping

In Unix, piping redirects the standard output of one command-line application into the standard input of another. It is represented by the '|' character, which combines two or more commands. The general structure is `cmd1 | cmd2 | cmd3 | .... | cmdN`, the standard output from `cmd1` is the standard input for `cmd2`, and so on.

Creating a simple command-line application that does one thing and one thing well follows the Unix philosophy. It reduces the complexity of a single CLI, so you'll see many examples of different applications that can be chained together by pipes. Here are a few examples:

- `cat file.txt | grep "word" | sort`
- `sort list.txt | uniq`

- find . -type f –name main.go | grep audio

As an example, let's create a command that takes in standard input from a common application. Let's call it `piper`:

```
cobra-cli add piper
piper created at /Users/marian/go/src/github.com/
  marianina8/application
```

For the newly generated `piper`Cmd's Run field, add the following lines:

```
reader := bufio.NewReader(os.Stdin)
s, _ := reader.ReadString('\n')
fmt.Printf("piped in: %s\n", s)
```

Now, compile and run the `piper` command with some piped-in input:

```
  echo "hello world" | ./application piper
piper called
piped in: hello world
```

Now, suppose your command has a standard output that is written to a broken pipe; the kernel will raise a `SIGPIPE` signal. This is received as input to the command-line application, which can then output an error regarding the broken pipe. Besides receiving signals from the kernel, other signals, such as `SIGINT`, can be triggered by users who press control character key combinations such as *Ctrl* + *C* that interrupt the application. This is just one type of signal and control character, but more will be discussed in the following section.

Signals and control characters

As the name implies, signals are another way to communicate specific and actionable input by signaling to a command-line application. Sometimes, these signals can be from the kernel, or from users that press control characters key combinations and trigger signals to the application. There are two different types of signals:

- **Synchronous signals** – triggered by errors that occur when the program executes. These signals include `SIGBUS`, `SIGFPE`, and `SIGSEGV`.

- **Asynchronous signals** – triggered from the kernel or another application. These signals include `SIGHUP`, `SIGINT`, `SIGQUIT`, and `SIGPIPE`.

> **Note**
>
> A few signals, such as SIGKILL and SIGSTOP, may not be caught by a program, so utilizing the os/signal package for custom handling will not affect the result.

There is a lot to discuss in depth on signals, but the main point is that they are just another method of receiving input. We'll stay focused on how this data is received by the command-line application. The following is a table explaining some of the most commonly used signals, control character combinations, and their descriptions:

Signal (Key Combination)	Description
SIGINT (Ctrl-C)	Interrupt application
SIGTSTP (Ctrl-Z)	Suspend application
SIGQUIT (Ctrl-\)	Quit application
SIGHUP	Hang up on controlling terminal or death of controlling process
SIGFPE	Illegal mathematical operation attempted
SIGKILL	Quit immediately with no cleanup process
SIGALRM	Alarm clock signal (for timers)
SIGTERM	Software termination signal (kill by default)

Figure 5.1 – Table of signals with related key combinations and descriptions

The following are two function calls added to rootCmd to handle exiting your application with grace when a SIGINT or SIGTSTP signal is received. The Execute function that calls rootCmd now looks like this:

```
func Execute() {
    SetupInterruptHandler()
    SetupStopHandler()
    err := rootCmd.Execute()
    if err != nil {
        os.Exit(1)
    }
}
```

The SetupInterruptHandler code is as follows:

```
func SetupInterruptHandler() {
    c := make(chan os.Signal)
    signal.Notify(c, os.Interrupt, syscall.SIGINT)
    go func() {
        <-c
```

```
        fmt.Println("\r- Wake up! Sleep has been
            interrupted.")
        os.Exit(0)
    }()
}
```

Similarly, the `SetupStopHandler` code is as follows:

```
func SetupStopHandler() {
    c := make(chan os.Signal)
    signal.Notify(c, os.Interrupt, syscall.SIGTSTP)
    go func() {
        <-c
        fmt.Println("\r- Wake up! Stopped sleeping.")
        os.Exit(0)
    }()
}
```

Now, we'll need a command to interrupt or stop the application. Let's use the Cobra CLI and add a `sleep` command:

```
  cobra-cli add sleep
sleep created at /Users/marian/go/src/github.com/
  marianina8/application
```

The Run field of `sleepCmd` is changed to run an infinite loop that prints out some Zs (`Zzz`) until a signal interrupts the `sleep` command and wakes it up:

```
Run: func(cmd *cobra.Command, args []string) {
    fmt.Println("sleep called")
    for {
        fmt.Println("Zzz")
        time.Sleep(time.Second)
    }
},
```

By running the `sleep` command and then using *Ctrl + C*, we get the following output:

```
  ./application sleep
sleep called
Zzz
Zzz
- Wake up!  Sleep has been interrupted.
- Wake up!  Stopped sleeping.
```

Trying again but now using *Ctrl + Z*, we get the following output:

```
  ./application sleep
sleep called
Zzz
Zzz
Zzz
- Wake up!  Stopped sleeping.
```

You can utilize signals to interrupt or quit your application gracefully or take action when an alarm is triggered. While commands, arguments, and flags are the most commonly known types of input for command-line applications, it is important to consider handling these signal inputs to create a more robust application. If a terminal hangs and `SIGHUP` is received, your application can save information on the last state and handle cleanup where necessary. In this case, while it's not as common, it's just as important.

User interaction

Although your user input can be in the form of commands, arguments, and flags, user interaction is more of a back-and-forth interaction between the user and the application. Suppose a user misses a required flag for a particular subcommand – your application can prompt the user and receive the value for that flag via standard input. Sometimes, rather than utilizing the more standard input of commands, arguments, and flags, an interactive command-line application can be built instead.

An interactive CLI would prompt for input and then receive it through `stdin`. There are some useful packages for building interactive and accessible prompts in Go. For the following examples, we'll use the `https://github.com/AlecAivazis/survey` package. There are multiple fun ways to prompt input using the `survey` package. A `survey` command will ask questions that need to be stored in a variable. Let's define it as qs, a slice of the `*survey.Question` type:

```
var qs = []*survey.Question{}
```

`survey` can prompt the user for different types of input, as defined here:

- **Simple text input**

    At a very basic level, users can receive basic text input:

    ```
    {
        Name: "firstname",
        Prompt: &survey.Input{Message: "What is your first
            name?"},
        Validate: survey.Required,
        Transform: survey.Title,
    },
    ```
    Output:
    ```
    ? What is your first name?
    ```

- **Suggesting options**

    This terminal option allows you to give the user suggestions for the prompted question:

    ```
    {
        Name: "favoritecolor",
        Prompt: &survey.Select{
        Message: "What's your favorite color?",
        Options: []string{"red", "orange", "yellow",
            "green", "blue", "purple", "black", "brown",
                "white"},
        Default: "white",
    },
    ```
    Output:
    ```
    ? What is your favorite color? [tab for suggestions]
    ```

Hitting the *Tab* key will show the available options:

```
? What is your favorite color? [Use arrows to
    navigate, enter to select, type to complement
        answer]
```

```
red
orange
yellow
green
blue
purple
black
brown
white
```

- **Inputting multiple lines**

When receiving input, sometimes, pressing the *Return* key will immediately pass any text received before directly as input to the program. Utilizing the `survey` package allows you to enter multiple lines before receiving input:

```
{
    Name: "story",
    Prompt: &survey.Multiline{
    Message: "Tell me a story.",
    },
},
Output:
    ? Tell me a story [Enter 2 empty lines to finish]
A long line time ago in a faraway town, there lived a
    princess who lived in a castle far away from the
        city.  She was always sleeping, until one day…
```

- **Protecting password input**

To keep data private, when inputting private information, the `survey` package will replace the characters with * symbols:

```
{
    Name: "secret",
```

```
    Prompt: &survey.Password{
    Message: "Tell me a secret",
    },
},
Output:
? Tell me a secret: ************
```

- **Confirming with Yes or No**

  Users can respond with a simple yes or no to the command prompt:

```
{
    Name: "good",
    Prompt: &survey.Confirm{
    Message: "Are you having a good day?",
    },
},
Output:
? Are you having a good day? (Y/n)
```

  Now, let us see how to select from a checkbox option.

- **Selecting from a checkbox option**

  Multiple options can be selected within a vertical checkbox option. Navigating the options is done with the up and down arrows, and selecting is done with the spacebar:

```
{
    Name: "favoritepies",
    Prompt: &survey.MultiSelect{
    Message: "What pies do you like:",
    Options: []string{"Pumpkin", "Lemon Meringue",
      "Cherry", "Apple", "Key Lime", "Pecan", "Boston
        Cream", "Rhubarb", "Blackberry"},
    },
},
Output:
? What pies do you like: [Use arrows to move, space to
select, <right> to all, <left> to none, type to
filter]
```

```
>  [ ]  Pumpkin
   [ ]  Lemon Meringue
   [ ]  Cherry
   [ ]  Apple
   [ ]  Key Lime
   [ ]  Pecan

... .
```

Create a new `survey` command with the following:

```
cobra-cli add survey
```

The Run field of `surveyCmd` creates a struct that receives all the answers to questions asked:

```
Run: func(cmd *cobra.Command, args []string) {
    fmt.Println("survey called")
    answers := struct {
        FirstName string
        FavoriteColor string
        Story string
        Secret string
        Good bool
        FavoritePies []string
    }{}
```

The `Ask` method then takes in the questions, `qs`, and then receives all the answers to the questions asked into a pointer to the `answers` struct:

```
    err := survey.Ask(qs, &answers)
    if err != nil {
        fmt.Println(err.Error())
        return
    }
```

Finally, the results are printed out:

```
    fmt.Println("********** SURVEY RESULTS **********")
    fmt.Printf("First Name: %s\n", answers.FirstName)
    fmt.Printf("Favorite Color: %s\n",
```

```
                  answers.FavoriteColor)
        fmt.Printf("Story: %s\n", answers.Story)
        fmt.Printf("Secret: %s\n", answers.Secret)
        fmt.Printf("It's a good day: %v\n", answers.Good)
        fmt.Printf("Favorite Pies: %s\n", answers.FavoritePies)
    },
```

Testing out the survey command, we get the following:

```
    ./application survey
survey called
? What is your first name? Marian
? What's your favorite color? white
? Tell me a story.
I went to the dodgers game last night and
they lost, but I still had fun!
? Tell me a secret ********
? Are you having a good day? Yes
? What pies do you prefer: Pumpkin, Lemon Meringue, Key
    Lime, Pecan, Boston Cream
********** SURVEY RESULTS **********
First Name: Marian
Favorite Color: white
Story: I went to the dodgers game last night and
they lost, but I still had fun!
Secret: a secret
It's a good day: true
Favorite Pies: [Pumpkin Lemon Meringue Key Lime Pecan
    Boston Cream]
```

Although these examples are just a selection of the many input prompts provided by the survey package, you can visit the GitHub page to view examples of all the possible options. Playing around with prompts reminds me of early text-based RPG games that used them to prompt the gamer's character. Having learned about the many different types of input, whether user-based, from the kernel, or from other piped applications, let's discuss how to process this incoming data.

Processing data

Data processing is when raw data is fed into a process, analyzed, and then used to generate useful information or output. At a very general level, this can include sorting data, searching or querying for data, and converting data from one type of input into another. For a CLI, the input can be received in the different ways discussed in the previous section. When receiving arguments using the Cobra framework, all the values are read in as string input. However, given a string of 123, we can do a type check by utilizing the strconv package's Atoi method, which converts an ASCII string into an integer:

```
val, err := strconv.Atoi("123")
```

If the string value cannot be converted because it isn't a string representation of an integer, then an error will be thrown. If the string is a representation of an integer, then the integer value will be stored in the val variable.

The strconv package can be used to check, with conversion, many other types, including Boolean, float, and uint values as well.

Flags, on the other hand, can have predefined types. Within the Cobra framework, the pflag package is used, which is just an extension of the standard go flag package. For example, when a flag is defined, you can define it specifically as a String, Bool, Int, or custom type. The preceding 123 value, if read in as an Int flag, could be defined with the following lines of code:

```
var intValue int
flag.IntVar(&intValue, "flagName", 123, "help message")
```

This can be done similarly for String and Bool flags. You can even create a flag with a custom, specific interface using the Var method:

```
var value Custom
flag.Var(&value, "name", "help message")
```

Just ensure that the Custom struct satisfies the following interface defined within the pflag package for custom flags:

```
// (The default value is represented as a string.)
type Value interface {
    String() string
    Set(string) error
    Type() string
}
```

I defined the `Custom` struct as the following:

```
type Custom struct {
    Value string
}
```

Therefore, the `Set` method is simply defined as follows:

```
func (c *Custom) Set(value string) error {
    c.Value = value
    return nil
}
```

Passing the value into the flag was handled by `flag: --name="custom value`. The `String` method is then used to print the value:

```
fmt.Println(cmd.Flag("name").Value.String())
```

It looks like this:

```
custom value
```

Besides passing in string values that can be converted into different types, oftentimes, a path to a file is passed in. There are multiple ways of reading data from files. Let's list each, along with a method to handle this way of reading in a file and a pro and a con for each:

- **In its entirety, all at once**: The `os.ReadFile` method reads the entire file and returns its contents. It does not error when encountering the **end of file (EOF)**:

```
func all(filename string) {
    content, err := os.ReadFile(filename)
    if err != nil {
        fmt.Printf("Error reading file: %s\n", err)
        return
    }
    fmt.Printf("content: %s\n", content)
}
```

- **Pros**: Faster performance

- **Cons**: Consumes more memory in a shorter amount of time

- **In predefined chunks**: The `file.Read` method reads in the buffer at its predetermined size and returns the bytes, which can be printed after being cast as a string. Unlike the `ioutil.ReadFile` method, `file.Read` from the buffer will error when it reaches the EOF:

```go
func chunk(file *os.File) {
    const size = 8 // chunk size
    buff := make([]byte, size)
    fmt.Println("content: ")
    for {
        // read content to buffer of size, 8 bytes
        read8Bytes, err := file.Read(buff)
        if err != nil {
            if err != io.EOF {
                fmt.Println(err)
            }
            break
        }
        // print content from buffer
        fmt.Println(string(buff[:read8Bytes]))
    }
}
```

- **Pros**: Easy to implement, consumes little memory

- **Cons**: If the chunks are not properly chosen, you may have inaccurate results, increased complexity when comparing or analyzing the data, and potential error propagation.

- **Line by line**: By default, a new scanner will split the text up by lines, so it's not necessary to define the `split` function. The `scanner.Text()` method reads into the next token that delimits each scan – in the following example, line by line. Finally, `scanner.Scan()` does not return an error when it encounters the EOF:

```go
func line(file *os.File) {
    scanner := bufio.NewScanner(file)
```

```
lineCount := 0
for scanner.Scan() {
    fmt.Printf("%d: %s\n", lineCount,
      scanner.Text())
    lineCount++
}
if err := scanner.Err(); err != nil {
    fmt.Printf("error scanning line by line:
        %s\n", err)
    }
  }
```

- **Pros**: Easy to implement – an intuitive way to read in data and output data.

- **Cons**: Processing an extremely large file may cause memory constraints. Increased complexity may cause inaccurate results, if the data is not well suited to line by line processing.

- **Word by word** To overwrite the default Split function, pass bufio.ScanWords into the Split function. This will then define the tokens between each word and scan between each token. Again, scanning in this way will not encounter an error at the EOF either:

```
func word(file *os.File) {
    scanner := bufio.NewScanner(file)
    scanner.Split(bufio.ScanWords)
    wordCount := 0
    for scanner.Scan() {
        fmt.Printf("%d: %s\n", wordCount,
            scanner.Text())
      wordCount++
    }
    if err := scanner.Err(); err != nil {
        fmt.Printf("error scanning by words: %s\n",
            err)
    }
  }
```

- **Pros**: Easy to implement – an intuitive way to read data and output data

- **Cons**: Inefficient and time consuming for large files. Increased complexity may cause inaccurate results, if the data is not well suited to word by word processing

Choosing the way to handle processing the data received from the file depends on the use case. Additionally, there are three main types of data processing: batch, online, and real-time.

As you can guess from the name, batch processing takes similar tasks that are collected, or batched, and then runs them simultaneously. Online processing requires internet connectivity to reach an API endpoint to fully process data and return a result. Real-time processing is the execution of data in such a short period that the data is instantaneously output.

Examples of different use cases requiring a specific type of processing vary. Bank transactions, billing, and reporting often use batch processing.

A CLI that utilizes an API behind the scenes would often require internet access to handle online processing. Real-time processing is used when timeliness is of utmost importance, often in manufacturing, fraud detection, and computer vision tools.

Once the data has been processed, the result must be returned to the user or receiving process. In the next section, we will discuss the details of returning the output and defining the best practices for returning data.

Returning the resulting output and defining best practices

When returning output from a process, it's important to know to who or what you're returning data. It's incredibly important to return output that's human-readable. However, to determine whether you're returning data to a human or a machine, check whether you're writing to a TTY. Remember TTY? You can refer to *Chapter 1, Understanding CLI Standards*, in which we discussed the history of the CLI interface and the term TTY, short for teletypewriter or teletype.

If writing to a TTY, we can check whether the stdout file descriptor refers to a terminal or not, and change the output depending on the result.

Let's check out this block of code, which checks whether the stdout file descriptor is writing to a TTY or not:

```
if fileInfo, _ := os.Stdout.Stat(); (fileInfo.Mode() &
    os.ModeCharDevice) != 0 {
    fmt.Println("terminal")
} else {
    fmt.Println("not a terminal")
}
```

Let's call it within the Run method of a command called tty using the following command:

```
./application tty
```

Then, the output is as follows,:

```
terminal
```

However, if we pipe the output to a file by calling `./application tty > file.txt`, then the contents of the file are as follows:

```
not a terminal
```

Certainly, it makes sense to add colored ASCII text when returning output to a human, but that's often useless and extraneous information for output to a machine process.

When writing output, always put humans first, specifically in terms of usability. However, if the machine-readable output does not affect usability, then output in machine-readable output. Because streams of text are universal input in Unix, it's typical for programs to be linked together by pipes. The output is typically lines of text, and programs expect input as lines of text as well. A user should expect to write output that can easily be grepped. You cannot know for sure where the output will be sent to and which other processes may be consuming the output. Always check whether the output is sent to a terminal and print for another program if it's not. However, if using a machine-readable output breaks usability, but the human-readable output cannot be easily processed by another machine process, default to human-readable output and then define the `-plain` flag to display this output as machine-readable output. Clean lines of text in tabular format are easily integrated with `grep` and `awk`. This gives the user the choice to define the format of the output.

Beyond defining the output for humans versus machines, it's standard to add a flag to define a specific format for the data returned. The `-json` flag is used when requesting data to be returned in JSON format and the `-xml` flag is used to request XML format. There's a Unix tool, `jq`, that can be integrated with a program's JSON output. In fact, this tool can manipulate any data returned in JSON format. Many tools within the Unix ecosystem take advantage of this and you can too.

Historically, because many of the older Unix programs were written for scripts or other programs, often, no output is returned on success. This can be confusing for users. Success cannot always be assumed, so it's ideal to display output on success. There's no reason to elaborate, so keep it brief and informative. Defining a `-quit` (or `-q`) flag can suppress unnecessary information if necessary.

Sometimes, a CLI can keep track of the state. The **GitHub CLI** is probably the best and most common example that many of you have already experienced. It does an excellent job of informing users of state changes and the current state using `git status`. This information needs to be transparent to the user, as it can often confirm the result of an action expected to change the state. The user understands their possible next steps by knowing the state.

Some of these next steps may also be suggested to the user. In fact, it's ideal to give users suggestions because it feels like they are being guided along, rather than left alone in the wild with a new CLI application. When a user first interacts with a CLI, it's best to make the learning experience similar to a guided adventure. Let's give a quick example in regard to GitHub's CLI. Consider when you have to merge the main branch into your current branch. Now and then, there'll be conflicts after the merge, and the CLI guides you when you check `git status`:

```
On branch {branch name}
Your branch and 'origin/{branch name}' have diverged
And have 1 and 1 different commits each, respectively.
    (use "git pull" to merge the remote branch into yours)
You have unmerged paths.
    (fix conflicts and run "git commit")
    (use "git merge -abort" to abort the merge)


Unmerged paths:
    (use "git add <file>..." to mark resolution)
            Both modified:      merge.json
```

> **Note**
>
> The response reminds the user of their current branch and state, as well as suggesting different options that the user could take. Not all CLIs handle the state, but when you do, it's best to make it well known and provide users with a clear path forward.

If there's any communication with a remote server, reading or writing of files (except for a cache), or any other actions that cross the boundary of the program's internals, communicate those actions to the user. I love HomeBrew's `install` command on their CLI. It's clear exactly what's going on behind the scenes when you use `brew install` for an application.

When a file is being downloaded or created, it's clearly stated:

```
==> Downloading https://ghcr.io/v2/homebrew/core/dav1d/
manifests/1.0.0
###################################################################
############### 100.0%
```

And look how hashtags are used to designate progress – they utilize ASCII characters in a way that increases information density. I love the cold glass of beer icon next to files existing in the `Cellar` folder. It makes you think of all the brew formulas existing inside a beer cellar. **Emojis** are worth a thousand words.

When an error is evoked, the text is displayed in red, intending to evoke a sense of urgency and alertness. Color, if used, must be used intentionally. A green failure, or red success, is confusing for users. I'm certain, just like utilizing ASCII art to increase information density, color has the same purpose. A green success cannot be mistaken easily for a failure, and vice versa. Make sure to make important information stand out by using colors infrequently. Too many colors will make it difficult for anything to stand out.

However, while color may excite some of us, it annoys others. There may be any number of reasons why someone may want to disable the color in their CLI. For whatever reason to continue in a black-and-white world, there are specific times color should not be used:

- When piping to another program
- When the NO_COLOR environment variable is set
- When the TERM environment variable is set to dumb
- When the -no-color flag is passed
- When your app's MYAPP_NO_COLOR environment variable is set

It goes without saying that if we don't allow colors, we don't allow animations either! Well, I won't tell you what to do, just try it for yourself – pipe an animation to a file via stdout. I dare you! You might end up with some great ASCII art, but it will be busy and difficult to understand the data. The goal is clarity. With ASCII art, color intent, and animations to increase the information density, we need to understand at some point that we need to use clear words that are understood by all. Consider your wording from the perspective of someone who is using your CLI for the first time. Guide users with your words.

As for printing log output, only do so under the verbose mode, represented by the -verbose flag and -v for short. Don't use the stderr file descriptor as a log file.

If a CLI outputs a lot of text at once, such as git diff, a pager is used. Thank goodness. This makes it so much easier to page through the output to review differences rather than receiving all the text at once. This is just one of the many ways that GitHub has delivered a very thoughtful CLI to its users.

Finally, make errors stand out – use red text or a red *x* emoji to increase understanding if an error occurs. If colors are disabled, then use text to communicate that an error has occurred and offer some suggestions for the next steps to take – and, even better, an avenue toward support via email or a website.

Summary

In this chapter, you learned about the command-line process – receiving input, processing data, and returning the output. The most popular different types of input have been discussed: from **subcommands**, **arguments**, and **flags**, to **signals** and **control characters**.

We created an interactive survey to receive input from a user and discussed data processing. We also learned how to take the first steps of processing: converting argument string data, converting and checking the type, receiving data from typed and custom flags, and finally, reading data from a file.

We also covered a brief explanation regarding the different types of processing: batch, online, and real-time processing. Ultimately, the use case will lead you to understand what sort of input you'll require, and whether running tasks in batches, over the internet, or in real time is required.

Returning the output is just as important as receiving it, if not more! This is your chance to create a more pleasant experience for your user. Now that you're developing for humans first, you have the opportunity to put yourself in their shoes.

How would you want to receive data in a way that makes you feel assured, understanding failures and what to do next, and where to find help? Not all processes run successfully, so let's at least make users feel that they're on the path to success. In *Part 2, Chapter 6, Calling External Processes, Handling Errors and Timeouts*, we will continue to discuss the command-line process in more detail, focusing on external processes and how to handle timeouts and errors and communicate them to the user effectively.

Questions

1. Are arguments or flags preferred for CLI programs? Why?
2. What key combination interrupts a computer process?
3. What flag can be added to your CLI to modify the output into plain output that can easily be integrated with tools such as `grep` and `awk`?

Answers

1. Flags are preferred for CLI programs because they make it much easier to add or remove functionality.
2. *Ctrl + C.*
3. The `-plain` flag can be added to remove any unnecessary data from the output.

Further reading

- What is a TTY? (`https://unix.stackexchange.com/questions/4126/what-is-the-exact-difference-between-a-terminal-a-shell-a-tty-and-a-con/4132#4132`)
- NO_COLOR (`https://no-color.org/`)
- *12 Factor CLI Apps* (`https://medium.com/@jdxcode/12-factor-cli-apps-dd3c227a0e46`)

Calling External Processes and Handling Errors and Timeouts

Many command-line applications interact with other external commands or API services. This chapter will guide you through how to call these external processes and how to handle timeouts and other errors when they occur. The chapter will start with a deep dive into the os/exec package, which contains everything you need to create commands that call external processes that give you multiple options for creating and running commands. You'll learn how to retrieve data from the standard output and standard error pipes, as well as creating additional file descriptors for similar usage.

Another external process involves calling external API service endpoints. The net/http package is discussed and is where we start defining the client, then create the requests that it executes. We will discuss the different ways requests can be both created and executed.

Timeouts and other errors can occur when calling either type of process. We will end the chapter by looking at how to capture when timeouts and errors occur in our code. It's important to be mindful that these things can happen and so it's important to write code that can handle them. The specific action taken upon error is dependent on the use case, so we'll discuss the code to capture these cases only. To summarize, we'll be covering the following topics:

- Calling external processes
- Interacting with REST APIs
- Handling the expected – timeouts and errors

Technical requirements

You'll need a UNIX operating system to understand and run the examples shared in the chapter.

You can also find the code examples on GitHub at https://github.com/PacktPublishing/Building-Modern-CLI-Applications-in-Go/tree/main/Chapter06.

Calling external processes

Within your command-line application, you may need to call some external processes. Sometimes, there are Golang libraries offered for third-party tools that function as a wrapper. For example, Go CV, `https://gocv.io/`, is a Golang wrapper offered for OpenCV, an open source computer vision library. Then, there's GoFFmpeg, `https://github.com/xfrr/goffmpeg`, which is a wrapper offered for FFmpeg, a library for recording, converting, and streaming audio and video files. Often, you need to install an underlying tool, such as OpenCV or FFmpeg, and then the library interacts with it. Calling these external processes then means importing the wrapper package and calling its methods within your code. Often, when you dive into the code, you'll find that these libraries provide a wrapper for the C code.

Besides importing a wrapper for an external tool, you may call external applications using the `os/exec` Golang library. This is the main purpose of the library and in this section, we will be digging into how to use it to call external applications.

First, let's review each of the variables, types, and functions that exist within the `os/exec` package with an example of each.

The os/exec package

By digging deeper into the `exec` package, you will find that it is a wrapper for the `os.StartProcess` method, making it easier to handle the remapping of standard in and standard out, connecting the input and output with pipes, and handling other modifications.

For clarity, it's important to note that this package does not invoke the operating system's shell and so doesn't handle tasks handled typically by the shell: expanding glob patterns, pipelines, or redirections. If it is necessary to expand glob patterns, then you can call the shell directly and make sure to escape values to make it safe, or you can also use the path or file path's `Glob` function. To expand any environment variables that exist in a string, use the `os` package's `ExpandEnv` function.

In the following subsections, we'll start to discuss the different variables, types, functions, and methods that exist within the `os/exec` package.

Variables

`ErrNotFound` is the error variable returned when an executable file is not found in the application's `$PATH` variables.

Types

Cmd is a struct that represents an external command. Defining a variable of this type is just in preparation for the command to be run. Once this variable, of the Cmd type, is run via either the Run, Output, or CombinedOutput method, it cannot be reused. There are several fields on this Cmd struct that we can also elaborate upon:

- Path string This is the only required field. It is the path of the command to run; if the path is relative, then it will be relative to the value stored in the Dir field.

- Args []string This field holds the arguments for the command. Args[0] represents the command. Path and Args are set when the command is run, but if Args is nil or empty, then just {Path} is used during execution.

- Env []string The Env field represents the environment for the command to run. Each value in the slice must be in the following format: "key=value". If the value is empty or nil, then the command uses the current environment. If the slice has duplicate key values, then the last value for the duplicate key is used.

- Dir string The Dir field represents the working directory of the command. If it's not set, then the current directory is used.

- Stdin io.Reader The Stdin field specifies the command process' standard input. If the data is nil, then the process reads from os.DevNull, the null device. However, if the standard input is *os.File, then the contents are piped. During execution, a goroutine reads from standard input and then sends that data to the command. The Wait method will not complete until the goroutine starts copying. If it does not complete, then it could be because of an **end-of-file** (**EOF**), read, or write-to-pipe error.

- Stdout io.Writer The Stdout field specifies the command process' standard output. If the standard output is nil, then the process connects to the os.DevNull null device. If the standard output is *os.File, then output is sent to it instead. During execution, a goroutine reads from the command process and sends data to the writer.

- `Stderr io.Writer` The `Stderr` field specifies the command process' standard error output. If the standard error is `nil`, then the process connects to the `os.DevNull` null device. If the standard error is `*os.File`, then error output is sent to it instead. During execution, a goroutine reads from the command process and sends data to the writer.

- `ExtraFiles []*os.File` The `ExtraFiles` field specifies additional files inherited by the command process. It doesn't include standard input, standard output, or standard error, so if not empty, entry x becomes the $3+x$ file descriptor. This field is not supported on Windows.

- `SysProcAttr *syscall.SysProcAttr SysProcAttr` holds system-specific attributes that are passed down to `os.StartProcess` as an `os.ProcAttr`'s `Sys` field.

- `Process *os.Process` The `Process` field holds the underlying process once the command is run.

- `ProcessState *os.ProcessState` The `ProcessState` field contains information about the process. It becomes available after the wait or run method is called.

Methods

The following are the methods that exist on the `exec.Cmd` object:

- `func (c *Cmd) CombinedOutput() ([]byte, error)` The `CombinedOutput` method returns both the standard output and standard error into 1-byte string output.

- `func (c *Cmd) Output ([]byte, error)` The `Output` method returns just the standard output. If an error occurs, it will usually be of the `*ExitError` type, and if the command's standard error, `c.Stderr`, is `nil`, `Output` populates `ExitError.Stderr`.

- `func (c *Cmd) Run() error` The Run method starts executing the command and then waits for it to complete. If there was no problem copying standard input, standard output, or standard error and the command exits with a zero status, then the error returned will be `nil`. If the command exits with an error, it will usually be of the `*ExitError` type, but could be other error types as well.

- `func (c *Cmd) Start() error`

- The `Start` method will start executing the command and not wait for it to complete. If the `Start` method runs successfully, then the `c.Process` field will be set. The `c.Wait` field will then return the exit code and release resources once complete.

- `func (c* Cmd) StderrPipe() (io.ReadCloser, error)` `StderrPipe` returns a pipe that is connected to the command's standard error. There won't be a need to ever close the pipe because the `Wait` method will close the pipe once the command exits. Do not call the `Wait` method until all reads from the standard error pipe have completed. Do not use this command with the `Run` method for the same reason.

- `func (c* Cmd) StdinPipe() (io.WriteCloser, error)` `StdinPipe` returns a pipe that is connected to the command's standard input. The pipe will be closed after `Wait`, and the command exits. However, sometimes the command will not run until the standard input pipe is closed, and thus you can call the `Close` method to close the pipe sooner.

- `func (c *Cmd) StdoutPipe() (io.ReadCloser, error)` The `StdoutPipe` method returns a pipe that is connected to the command's standard output. There's no need to close the pipe because `Wait` will close the pipe once the command exits. Again, do not call `Wait` until all reads from the standard output pipe have completed. Do not use this command with the `Run` method for the same reason.

- `func (c *Cmd) String() string` The `String` method returns a human-readable description of the command, c, for debugging purposes. The specific output may differ between Go version releases. Also, do not use this as input to a shell, as it's not suitable for that purpose.

- `func (c *Cmd) Wait() error` The `Wait` method waits for any copying to standard input, for standard output or standard error to complete, and for the command to exit. To utilize the `Wait` method, the command must have been started by the `Start` method and not the `Run` method. If there are no errors with copying from pipes and the process exits with a 0 exit status code, then the error returned will be `nil`. If the command's `Stdin`, `Stdout`, or `Stderr` field is not set to `*os.File`, then `Wait` also ensures that the respective input-output loop process completes as well.

`Error` is a struct that represents an error returned from the `LookPath` function when it fails to recognize the file as an executable. There are a couple of fields and methods of this specific error type that we will define in detail.

The following are the methods that exist on the `Error` type:

- `func (e *Error) Unwrap() error` If the error returned is a chain of errors, then you can utilize the `Unwrap` method to *unwrap* it and determine what kind of error it is.

`ExitError` is a struct that represents an error when a command exits unsuccessfully. `*os.ProcessState` is embedded into this struct, so all values and fields will also be available to the `ExitError` type. Finally, there are a few fields of this type that we can define in more detail:

- `Stderr []byte` This field holds a set of the standard error output responses if not collected from the `Cmd.Output` method. `Stderr` may only contain the prefix and suffix of the error output if it's sufficiently long. The middle will contain text about the number of omitted bytes. For debugging purposes, and if you want to include the entirety of the error messages, then redirect to `Cmd.Stderr`.

The following is the method that exists on the `ExitError` type:

- `func (e *ExitError) Error() string` The `Error` method returns the exit error represented as a string.

Functions

The following are functions that exist within the `os/exec` package:

- `func LookPath(file string) (string, error)` The `LookPath` function checks to see whether the file is an executable and can be found. If the file is a relative path, then it is relative to the current directory.

- `func Command(name string, arg ...string) *Cmd` The `Command` function returns the `Cmd` struct with just the path and args set. If name has path separators, then the `LookPath` function is used to confirm the file is found and executable. Otherwise, `name` is used directly as the path. This function behaves slightly differently on Windows. For example, it will execute the whole command line as a single string, including quoted args, then handle its own parsing.

- `func CommandContext(ctx context.Context, name string, arg ...string) *Cmd` Similar to the `Command` function, but receives context. If the context is executed before the command completes, then it will kill the process by calling `os.Process.Kill`.

Now that we've really dived deep into the os/exec package and the structs, functions, and methods needed to execute functions, let's actually use them in code to execute a function externally. Let's create commands using the Cmd struct, but also with the Command and CommandContext functions. We can then take one example command and run it using either the Run, Output, or CombinedOutput method. Finally, we will handle some errors typically returned from these methods.

> **Note**
>
> If you want to follow along with the examples coming up, within the Chapter-6 repository, install the necessary applications. In Windows, use the .\build-windows.p1 PowerShell script. In Darwin, use the make install command. Once the applications are installed, run go run main.go.

Creating commands using the Cmd struct

There are several different ways of creating commands. The first way is with the Cmd struct within the exec package.

Using the Cmd struct

We first define the cmd variable with an unset Cmd structure. The following code resides in /examples/command.go within the CreateCommandUsingStruct function:

```
cmd := exec.Cmd{}
```

Each field is set separately. The path is set using filepath.Join, which is safe for use across different operating systems:

```
cmd.Path = filepath.Join(os.Getenv("GOPATH"), "bin",
"uppercase")
```

Each field is set separately. The Args field contains the command name in the Args[0] position, followed by the rest of the arguments to be passed in:

```
cmd.Args = []string{"uppercase", "hack the planet"}
```

The following three file descriptors are set – Stdin, Stdout, and Stderr:

```
cmd.Stdin = os.Stdin // io.Reader
cmd.Stdout = os.Stdout // io.Writer
cmd.Stderr = os.Stderr // io.Writer
```

However, there's a `writer`, file descriptor that's passed into the `ExtraFiles` field. This specific field is inherited by the command process. It's important to note that a pipe won't work if you don't pass the writer in `ExtraFiles`, because the child must get the writer to be able to write to it:

```
reader, writer, err := os.Pipe()
if err != nil {
    panic(err)
}
cmd.ExtraFiles = []*os.File{writer}
if err := cmd.Start(); err != nil {
    panic(err)
}
```

Within the actual uppercase command that's called, there's code in `cmd/uppercase/uppercase.go` that takes the first argument after the command name and changes the case to uppercase. The new uppercased text is then encoded into the pipe or extra file descriptor:

```
input := os.Args[1:]
output := strings.ToUpper(strings.Join(input, ""))
pipe := os.NewFile(uintptr(3), "pipe")
err := json.NewEncoder(pipe).Encode(output)
if err != nil {
    panic(err)
}
```

Back to the `CreateCommandUsingStruct` function, the value that's encoded into the pipe can now be read via the `read` file descriptor of the pipe and then output with the following code:

```
var data string
decoder := json.NewDecoder(reader)
if err := decoder.Decode(&data); err != nil {
    panic(err)
}
fmt.Println(data)
```

We now know one way of creating a command using the `Cmd` struct. Everything could have been defined at once at the same time as the command was initialized and depends on your preference.

Using the Command function

Another way to create a command is with the exec.Command function. The following code resides in /examples/command.go within CreateCommandUsingCommandFunction:

```
cmd := exec.Command(filepath.Join(os.Getenv("GOPATH"), "bin",
"uppercase"), "hello world")
reader, writer, err := os.Pipe()
if err != nil {
    panic(err)
}
```

The exec.Command function takes the file path to the command as the first argument. A slice of strings representing the arguments is optionally passed for the remaining parameters. The rest of the function is the same. Because exec.Command does not take any additional parameters, we similarly define the ExtraFiles field outside the original variable initialization.

Running the command

Now that we know how to create commands, there are multiple different ways to run or start running a command. While each of these methods has already been described in detail earlier in this section, we'll now share an example of using each.

Using the Run method

The Run method, as mentioned earlier, starts the command process, and then waits for its completion. The code for this is called from the main.go file but can be found under /examples/running.go. In this example, we call a different command called lettercount, which counts the letters in a string and then prints out the result:

```
cmd := exec.Command(filepath.Join(os.Getenv("GOPATH"), "bin",
"lettercount"), "four")
cmd.Stdin = os.Stdin
cmd.Stdout = os.Stdout
cmd.Stderr = os.Stderr
var count int
```

Again, we use the ExtraFiles field to pass in an additional file descriptor to write the result to:

```
reader, writer, err := os.Pipe()
if err != nil {
    panic(err)
```

```
    }
    cmd.ExtraFiles = []*os.File{writer}
    if err := cmd.Run(); err != nil {
        panic(err)
    }
    if err := json.NewDecoder(reader).Decode(&count); err != nil {
        panic(err)
    }
```

The result is finally printed with the following code:

```
    fmt.Println("letter count: ", count)
```

Using the Start command

The `Start` method is like the `Run` method; however, it doesn't wait for the process to complete. You can find the code that uses the `Start` command in `examples/running.go`. For the most part, it's identical, but you'll be replacing the code block containing `cmd.Run` with the following:

```
    if err := cmd.Start(); err != nil {
        panic(err)
    }
    err = cmd.Wait()
    if err != nil {
        panic(err)
    }
```

It's very important to call the `cmd.Wait` method because it releases resources taken by the command process.

Using the Output command

As the method name suggests, the `Output` method returns anything that's been piped into the standard out pipe. The most common way to push from a command to the standard output pipe is through any of the print methods in the `fmt` package. An additional line is added to the end of the main function for the `lettercount` command:

```
    fmt.Printf("successfully counted the letters of \"%v\" as
    %d\n", input, len(runes))
```

The only difference within the code that utilizes this `Output` method, which can be found in the `examples/running.go` file under the `OutputMethod` function, is this line of code:

```
out, err := cmd.Output()
```

The `out` variable is a byte slice that can later be cast to a string to be printed out. This variable captures the standard out and when the function is run, the output displayed is as follows:

output: successfully counted the letters of "four" as 4

Using the CombinedOutput command

As the method name suggests, the `CombinedOutput` method returns a combined output of the standard output and standard error piped data. Add a line toward the end of the `lettercount` command's `main` function:

```
fmt.Fprintln(os.Stderr, "this is where the errors go")
```

The only big difference between the calls from the previous function and the current function, `CombinedOutputMethod`, is this line:

```
CombinedOutput, err := cmd.CombinedOutput()
```

Similarly, it returns a byte slice, but now contains the combined output of standard error and standard output.

Executing commands on Windows

Alongside the examples are similar files that end with `_windows.go`. The major thing to note, in the previous examples, is that `ExtraFiles` is not supported on Windows. These Windows-specific and simple examples execute an external `ping` command to `google.com`. Let's take a look at one:

```
func CreateCommandUsingCommandFunction() {
    cmd := exec.Command("cmd", "/C", "ping", "google.com")
    output, err := cmd.CombinedOutput()
    if err != nil {
        panic(err)
    }
    fmt.Println(string(output))
}
```

Like the commands we've written for Darwin, we can create commands using the `exec.Command` function or the struct and call `Run`, `Start`, `Wait`, `Output`, and `CombinedOutput` just the same.

Also, for pagination, `less` is used on Linux and UNIX machines, but `more` is used on Windows. Let's quickly show this code:

```
func Pagination() {
    moreCmd := exec.Command("cmd", "/C", "more")
    moreCmd.Stdin = strings.NewReader(blob)
    moreCmd.Stdout = os.Stdout
    moreCmd.Stderr = os.Stderr
    err := moreCmd.Run()
    if err != nil {
        panic(err)
    }
}
var (
    blob = `
    ...
    `
)
```

Similarly, we can pass in the name and all arguments using the `exec.Command` method. We also pass the long text into the `moreCmd.Stdin` field.

So, the `os/exec` package offers different ways to create and run external commands. Whether you create a quick command using the `exec.Command` method or directly create one with the `exec.Cmd` struct and then run the `Start` command, you have options. Finally, you can either retrieve the standard output and error output separately or together. Knowing all about the `os/exec` package will make it easy to successfully run external commands from your Go command-line application.

Interacting with REST APIs

Often, if a company or user has already created an API, the command-line application will send requests to either the REST API or the gRPC endpoints. Let's first talk about using REST API endpoints. It is important to understand the `net/http` package. It's quite a large package with many types, methods, and functions, many of which are used for development on the server side. In this context, the command-line application will be the client of the API, so we won't discuss each in detail. We'll go into a few basic use cases from the client side though.

Get request

Let's revisit the code from *Chapter 3, Building an Audio Metadata CLI*. Within the Run command of the CLI command code, found in the /cmd/cli/command/get.go file, is a snippet of code that calls the corresponding API request endpoint using the GET method:

```
params := "id=" + url.QueryEscape(cmd.id)
path := fmt.Sprintf("http://localhost/request?%s", params)
payload := &bytes.Buffer{}
method := "GET"
client := cmd.client
```

Notice that in the preceding code, we take the field value, id, which has been set on the cmd variable, and pass it into the HTTP request as a parameter. Consider the flags and arguments to be passed which are to be used as parameters for your HTTP request. The following code executes the request:

```
req, err := http.NewRequest(method, path, payload)
if err != nil {
    return err
}
resp, err := client.Do(req)
if err != nil {
    return err
}
defer resp.Body.Close()
```

Finally, the response is read into a byte string and printed. Prior to accessing the body of the response, check whether the response or body is nil. This can save you from some future headaches:

```
b, err := io.ReadAll(resp.Body)
if err != nil {
    return err
}
fmt.Println(string(b))
return nil
```

However, in reality, there will be much more done with the response:

1. **Check the response status code**: If the response is 200 OK, then we can return the output as it was a successful response. Otherwise, in the next section, *Handling the expected – timeouts and errors*, we'll discuss how to handle other responses.

2. **Log the response**: We may, ideally, log the response if it doesn't contain any sensitive data. This detailed information can be written to a log file or output when in verbose mode.

3. **Store the response**: Sometimes, the response may be stored in a local database or cache.

4. **Transform the data**: This returned data may also be unmarshaled into a local data struct. The struct types of data returned must be defined and, preferably, would utilize the same struct models defined within the API. In that case, if `Content-Type` in the header is set to `application/json`, we would unmarshal the JSON response into the struct.

Currently, in the audiofile application, we transform the data into an `Audio` struct like this:

```
var audio Audio
If err := json.Unmarshal(b, &audio); err != nil {
    fmt.Println("error unmarshalling JSON response"
}
```

But what if the response body isn't in JSON format and the content type is something else? In a perfect world, we'd have API documentation that informs us of what to expect so we can handle it accordingly. Alternatively, you can check to confirm the type first using the following:

```
contentType := http.DetectContentType(b) // b are the bytes
from reading the resp.Body
```

A quick search on the internet for HTTP content types will return a large list. In the preceding example, the audio company might have decided to return a `Content-Type` value of `audio/wave`. In that case, we could either download or stream the result. There are also different HTTP method types defined as constants within the `net/http` package:

- `MethodGet`: Used when requesting data

- `MethodPost`: Used for inserting data

- `MethodPut`: Request is idempotent, used for inserting or updating an entire resource

- `MethodPatch`: Similar to `MethodPut`, but sends only partial data to update without modifying the entire resource

- `MethodDelete`: Used for deleting or removing data

- `MethodConnect`: Used when talking to a proxy, when the URI begins with `https://`

- `MethodOptions`: Used to describe the communication options, or allowed methods, with the target

- `MethodTrace`: Used for debugging by providing a message loop-back along the path of the target

There are many possibilities for the method types and content types of data returned. In the preceding `Get` example, we use a client's `Do` method to call the method. Another option is to use the `http.Get` method. If we use that method, then we would use this code instead to execute the request:

```
resp, err := http.Get(path)
if err != nil {
    return err
}
defer resp.Body.Close()
```

Similarly, rather than using the `client.Do` method for a post or to post a form, there are specific `http.Post` and `http.PostForm` methods that can be used instead. There are times when one method works better for what you are doing. At this point, it's just important to understand your options.

Pagination

Suppose there is a large amount of data being returned by the request. Rather than overloading the client by receiving the data all at once, often pagination is an option. There are two fields that can be passed in as parameters to the call:

- `Limit`: The number of objects to be returned
- `Page`: The cursor for multiple pages of results returned

We can define these internally and then formulate the path as follows:

```
path := fmt.Sprintf("http://localhost/
request?limit=%d&page=%d", limit, page)
```

Make sure, if you're using an external API, to construct their documentation with the proper parameters for pagination and usage. This is just a general example. In fact, there are several other ways of doing pagination. You can send additional requests in a loop, incrementing the page until all data is retrieved.

From the command side, however, you could return all the data after pagination, but you can also handle pagination on the CLI side. A way to handle it on the client side after a large amount of data is collected from an HTTP `Get` request is to pipe the data. This data can be piped into the operating system's pager command. For UNIX, `less` is the pager command. We create the command and then pipe the string output to the `Stdin` pipe. This code can be found in the `examples/pagination.go` file. Similar to the other examples we've shared when creating a command, we create a pipe and pass in the writer as an extra file descriptor to the command so that data may be written out:

```go
pagesCmd := exec.Command(filepath.Join(os.Getenv("GOPATH"),
"bin", "pages"))
reader, writer, err := os.Pipe()
if err != nil {
    panic(err)
}
pagesCmd.Stdin = os.Stdin
pagesCmd.Stdout = os.Stdout
pagesCmd.Stderr = os.Stderr
pagesCmd.ExtraFiles = []*os.File{writer}
if err := pagesCmd.Run(); err != nil {
    panic(err)
}
```

Again, the data from the reader is decoded into the `data string` variable:

```go
var data string
decoder := json.NewDecoder(reader)
if err := decoder.Decode(&data); err != nil {
    panic(err)
}
```

This string is then passed into the `Strings.NewReader` method and defined as the input for the `less` UNIX command:

```go
lessCmd := exec.Command("/usr/bin/less")
lessCmd.Stdin = strings.NewReader(data)
lessCmd.Stdout = os.Stdout
err = lessCmd.Run()
if err != nil {
    panic(err)
}
```

When the command is run, the data is output as pages. The user then can press the spacebar to continue to the next page or use any of the command keys to navigate the data output.

Rate limiting

Often, when dealing with third-party APIs, there's a limit to how many requests can be handled within a particular time. This is commonly known as **rate limiting**. For a single command, you might require multiple requests to an HTTP endpoint and so you might prefer to limit how often you're sending these requests. Most public APIs will inform users of their rate limits, but there are times when you'll hit the rate limit of an API unexpectedly. We'll discuss how to limit your requests to stay within the limits.

There is a useful library, `x/time/rate`, that can be used to define the limit, which is how often something should be executed, and limiters that control the process from executing within the limit. Let's use some example code, supposing we want to execute something every five seconds.

The code for this particular example is located in the `examples/limiting.go` file. To reiterate, this is just an example and there are different ways to use `runner`. We're going to cover just a basic use case. We start by defining a struct that contains a function, Run, and the `limiter` field, which controls how often it will run. The `Limit()` function will use the `runner` struct to call a function within a rate limit:

```
type runner struct {
    Run func() bool
    limiter *rate.Limiter
}
func Limit() {
    thing := runner{}
    start := time.Now()
```

After defining `thing` as a `runner` instance, we get the start time and then define the function of `thing`. If the call is allowed within the time, because it does not exceed the limit, we print the current timestamp and return a `false` variable. We exit the function when at least 30 seconds have passed:

```
    thing.Run = func() bool {
        if thing.limiter.Allow() {
            fmt.Println(time.Now()) // or call request
            return false
        }
        if time.Since(start) > 30*time.Second {
            return true
        }
```

```
        return false
    }
```

We define the limiter for `thing`. We've used a customer variable, which we'll look at in more detail shortly. Simply, the `NewLimiter` method takes two variables. The first parameter is the limit, one event every five seconds, and the second parameter allows bursts for, at most, a single token:

```
    thing.limiter = rate.NewLimiter(forEvery(1, 5*time.
    Second),       1)
```

For those not familiar with the difference between a limit and a burst, a burst defines the number of concurrent requests the API can handle. The rate limit is the number of requests allowed per the defined time.

Next, inside a `for` loop, we call the Run function and only break when it returns `true`, which should be after 30 seconds have passed:

```
    for {
        if thing.Run() {
            break
        }
    }
}
```

As mentioned, the `forEvery` function, which returns a rate limit, is passed into the `NewLimiter` method. It simply calls the `rate.Every` method, which takes the minimum time interval between events and converts it into a limit:

```
func forEvery(eventCount int, duration time.Duration) rate.
Limit {
    return rate.Every(duration / time.Duration(eventCount))
}
```

We run this code and the timestamps are output. Notice that they are output every five seconds:

```
2022-09-11 18:45:44.356917 -0700 PDT m=+0.000891459
2022-09-11 18:45:49.356877 -0700 PDT m=+5.000891042
2022-09-11 18:45:54.356837 -0700 PDT m=+10.000891084
2022-09-11 18:45:59.356797 -0700 PDT m=+15.000891084
2022-09-11 18:46:04.356757 -0700 PDT m=+20.000891167
2022-09-11 18:46:09.356718 -0700 PDT m=+25.000891167
```

There are other ways of handling limiting requests, such as using a `time.Sleep(d Duration)` method after the code that is called inside a loop. I suggest using the `rate` package because it is great for not only limiting executions but also handling bursts. It has a lot more functionality that can be used for more complex situations when you are sending requests to an external API.

You've now learned how to send requests to external APIs and how to handle the response, and when you receive a successful response, how to transform and paginate the results. Also, because rate limiting is commonly required for APIs, we've discussed how to do that. Since this section has only handled the case of success, let's consider how to handle the case of failure in the following section.

Handling the expected – timeouts and errors

When building a CLI that calls external commands or sends HTTP requests to an external API, with data that is passed in by the user, it's a good idea to expect the unexpected. In a perfect world, you can guard against bad data. I'm sure you are familiar with the phrase *garbage in, garbage out*. You can create tests that also ensure that your code is covered for as many bad cases as you can think of. However, timeouts and errors happen. It's the nature of software, and as you come across them within your development and also in production, you can modify your code to handle new cases.

Timeouts with external command processes

Let's first discuss how to handle timeouts when calling external commands. The timeout code exists within the `examples/timeout.go` file. The following is the entire method, which calls the `timeout` command. If you take a look at the `timeout` command code, located within `cmd/timeout/timeout.go`, you'll see that it contains a basic infinite loop. This command will time out, but we need to handle the timeout with the following code:

```go
func Timeout() {
    errChan := make(chan error, 1)

    cmd := exec.Command(filepath.Join(os.Getenv("GOPATH"),
            "bin", "timeout"))
    if err := cmd.Start(); err != nil {
        panic(err)
    }
    go func() {
        errChan <- cmd.Wait()
    }()
    select {
        case <-time.After(time.Second * 10):
            fmt.Println("timeout command timed out")
```

```
            return
        case err := <-errChan:
            if err != nil {
                fmt.Println("timeout error:", err)
            }
        }
    }
}
```

We first define an error channel, errChan, which will receive any error returned from the cmd. Wait() method. The command, cmd, is then defined, and next cmd's Start method is called to initiate the external process. Within a Go function, we wait for the command to return using the cmd.Wait() method. errChan will only receive the error value once the command has exited and the copying to standard input and standard error has completed. Within the following select block, we wait to receive from two different channels. The first case waits for the time returned after 10 seconds. The second case waits for the command to complete and receive the error value. This code allows us to gracefully handle any timeout issues.

Errors or panics with external command processes

First, let's define the difference between errors and panics. Errors occur when the application can be recovered but is in an abnormal state. If a panic occurs, then something unexpected happened. For example, we try to access a field on a nil pointer or attempt to access an index that is out of bounds for an array. We can start by handling errors.

There are a couple of errors that exist within the os/exec package:

- exec.ErrDot: Error when the file path of the command failed to resolve within the current directory, ".", hence the name ErrDot

- exec.ErrNotFound: Error when the executable fails to resolve in the defined file path

You can check for the type to handle each error uniquely.

Handling errors when a command's path cannot be found

The following code exists within the examples/error.go file in the HandlingDoesNotExistErrors function:

```
cmd := exec.Command("doesnotexist", "arg1")
if errors.Is(cmd.Err, exec.ErrDot) {
    fmt.Println("path lookup resolved to a local directory")
}
if err := cmd.Run(); err != nil {
```

```
    if errors.Is(err, exec.ErrNotFound) {
        fmt.Println("executable failed to resolve")
    }
}
```

When checking the type of the command, use the `errors.Is` method, rather than checking whether `cmd.Err == exec.ErrDot` because the error is not returned directly. The `errors.Is` method checks the error chain for any occurrence of the specific error type.

Handling other errors

Also, within the `examples/error.go` file is handling an error thrown by the command process itself. This second method, `HandlingOtherMethods`, sets the command's standard error to a buffer that we can later use if an error is returned from the command. Let's take a look at the code:

```
cmd := exec.Command(filepath.Join(os.Getenv("GOPATH"), "bin",
"error"))
var out bytes.Buffer
var stderr bytes.Buffer
cmd.Stdout = &out
cmd.Stderr = &stderr
if err := cmd.Run(); err != nil {
    fmt.Println(fmt.Sprint(err) + ": " + stderr.String())
    return
}
fmt.Println(out.String())
```

When an error is encountered, we print not only the error, `exit status 1`, but also any data that has been piped into the standard error pipe, which should give the users more detail on why the error occurred.

To further understand how this code works, let's take a look at the error command implementation that exists in the `cmd/error/error.go` file:

```
func main() {
    if len(os.Args) != 0 { // not passing in any arguments in
this example throws an error
        fmt.Fprintf(os.Stderr, "missing arguments\n")
        os.Exit(1)
    }
```

```
        fmt.Println("executing command with no errors")
    }
```

Since we are not passing any arguments into the command function, after we check the length of `os.Args`, we print to the standard error pipe the reason we are exiting with a non-zero exit code. This is a very simple way to handle errors in an effective manner. When calling this external process, we just return the errors, but as we've all probably experienced, error messages can be a bit cryptic. In later chapters, we will talk about how we can rewrite these to be more human-readable and provide a few examples.

In *Chapter 4*, *Popular Frameworks for Building CLIs*, we discussed the use of the RunE function within the Cobra Command struct, which allows us to return an error value when the command is run. If you are calling an external process within the RunE method, then you can capture and return the error to the user, after rewriting it to a more human-readable format, of course!

Panics are handled differently than errors, but it is a good practice to, within your own code, provide a way to recover from a panic gracefully. You can see this code initiated within the `examples/panic.go` file within the `Panic` method. This calls the `panic` command, located in `cmd/panic/panic.go`. This command simply panics and then recovers. It returns the panic message to the standard error pipe, prints the stack, and exits with a non-zero exit code:

```
defer func() {
    if panicMessage := recover(); panicMessage != nil {
        fmt.Fprintf(os.Stderr, "(panic) : %v\n", panicMessage)
        debug.PrintStack()
        os.Exit(1)
    }
}()
panic("help!")
```

On the side that runs this command, we handle it just like any other error by capturing the error and printing data piped into the standard error.

Timeouts and other errors with HTTP requests

Similarly, you could also experience errors when sending requests to an external API server. To be clear, timeouts are considered errors as well. The code for this example is located within `examples/http.go`, which contains two functions:

- `HTTPTimeout()`
- `HTTPError()`

Before we dig into the previous methods, let's talk about the code that needs to be running in order for these methods to execute properly.

The cmd/api/ folder contains the code for defining the handlers and starting an HTTP server locally. The mux.HandleFunc method defines the request pattern and matches it to the handler function. The server is defined by its address, which runs on localhost, port 8080, and the Handler, mux. Finally, the server.ListenAndServe() method is called on the defined server:

```go
func main() {
    mux := http.NewServeMux()
    server := &http.Server{
        Addr: ":8080",
        Handler: mux,
    }

    mux.HandleFunc("/timeout", timeoutHandler)
    mux.HandleFunc("/error", errorHandler)

    err := server.ListenAndServe()
    if err != nil {
        fmt.Println("error starting api: ", err)
        os.Exit(1)
    }
}
```

The timeout handler is defined simply. It waits two seconds before sending the response by using the time.After(time.Second*2) channel:

```go
func timeoutHandler(w http.ResponseWriter, r *http.Request) {
    fmt.Println("got /timeout request")
    <-time.After(time.Second * 2)
    w.WriteHeader(http.StatusOK)
    w.Write([]byte("this took a long time"))
}
```

The error handler returns a status code of http.StatusInternalServerError:

```go
func errorHandler(w http.ResponseWriter, r *http.Request) {
    fmt.Println("got /error request")
    w.WriteHeader(http.StatusInternalServerError)
```

```
       w.Write([]byte("internal service error"))
   }
}
```

In a separate terminal, run the `make install` command inside the root of the repository to start the API server. Now, let's look at the code that calls each endpoint and show how we handle it. Let's first discuss the first type of error – the timeout:

- HTTPTimeout: Inside the `examples/http.go` file resides the `HTTPTimeout` method. Let's walk through the code together:

  - First, we *define the client* using the `http.Client` struct, specifying the timeout as one second. Remember that as the timeout handler on the API returns a response after two seconds, the request is sure to timeout:

```
client := http.Client{
    Timeout: 1 * time.Second,
}
```

  - Next, we *define the request*: a GET method to the `/timeout` endpoint. We pass in an empty body:

```
body := &bytes.Buffer{}
req, err := http.NewRequest(http.MethodGet, "http://
localhost:8080/timeout", body)
if err != nil {
    panic(err)
}
```

  - The client Do method is called with the request variable passed in as a parameter. We wait for the server to respond within a second and if not, an error is returned. Any errors returned by the client's Do method will be of the `*url.Error` type. You can access the different fields to this error type, but in the following code, we check whether the error's `Timeout` method returns `true`. In this statement, we can act however we'd like. We can return the error for now. We can back off and retry or we can exit. It depends on what your specific use case is:

```
resp, err := client.Do(req)
if err != nil {
    urlErr := err.(*url.Error)
    if urlErr.Timeout() {
        fmt.Println("timeout: ", err)
        return
    }
```

```
}
defer resp.Body.Close()
```

When this method is executed, the output is printed:

```
timeout:  Get "http://localhost:8080/timeout": context deadline
exceeded (Client.Timeout exceeded while awaiting headers)
```

A timeout is just one error, but there are many others you might encounter. Since the client Do method returns a particular error type in the net/url package, let's discuss that. Inside the net/url package exists the url.Error type definition:

```
type Error struct {
    Op  string // Operation
    URL string // URL
    Err error // Error
}
```

The error contains the Timeout() method, which returns true when a request times out, and it is important to note that when the response status is anything other than 200 OK, the error is not set. However, the status code indicates an error response. Error responses can be split into two different categories:

- **Client error responses** (status codes range from 400 to 499) indicate an error on the client's side. A few examples of this include Bad Request (400), Unauthorized (401), and Not Found (404).

- **Server error messages** (status codes range from 500 to 599) indicate an error on the server side. A few common examples of this include Internal Server Error (500), Bad Gateway (502), and Service Unavailable (503).

HTTPErrors: Some sample code of how this can be handled exists within the examples/http.go file within the HTTPErrors method. Again, it's important to make sure that the API server is running before executing this code:

- The code within the method starts by calling a GET request to the /error endpoint:

```
resp, err := http.Get("http://localhost:8080/error")
```

- If the error is not `nil`, then we cast it to the `url.Error` type to access the fields and methods within it. For example, we check whether `urlError` is a timeout or a temporary network error. If it is neither, then we can output as much information as we know about the error to standard output. This additional information can help us to determine what steps to take next:

```
if err != nil {
    urlErr := err.(*url.Error)
    if urlErr.Timeout() {
        // a timeout is a type of error
        fmt.Println("timeout: ", err)
        return
    }
    if urlErr.Temporary() {
        // a temporary network error, retry later
        fmt.Println("temporary: ", err)
        return
    }
    fmt.Printf("operation: %s, url: %s, error: %s\n", urlErr.
        Op,          urlErr.URL, urlErr.Error())
    return
}
```

- Since the status code error response isn't considered a Golang error, the response body might have some useful information. If it's not `nil`, then we can read the status code:

```
if resp != nil {
    defer resp.Body.Close()
```

- We initially check that `StatusCode` doesn't equal `http.StatusOK`. From there, we can check for particular error messages and take the appropriate action. In this example, we only check for three different types of error responses, but you can check for whichever ones make sense for what you're doing:

```
if resp.StatusCode != http.StatusOK {
    // action for when status code is not okay
    switch resp.StatusCode {
    case http.StatusBadRequest:
        fmt.Printf("bad request: %v\n", resp.Status)
    case http.StatusInternalServerError:
```

```
        fmt.Printf("internal service error: %v\n", resp.
            Status)
    default:
        fmt.Printf("unexpected status code: %v\n", resp.
            StatusCode)
    }
}
```

- Finally, a client or server error status does not necessarily mean that the response body is `nil`. We can output the response body in case there's any useful information we can further gather:

```
    data, err := ioutil.ReadAll(resp.Body)
    if err != nil {
        fmt.Println("err:", err)
    }
    fmt.Println("response body:", string(data))
}
```

This concludes the section for handling HTTP timeouts and other errors. Although the examples are simple, they give you the necessary information and guidance to handle timeouts, temporary networks, and other errors.

Summary

Over the course of this chapter, you've learned about the os/exec package in depth. This included learning about the different ways to create commands: using the command struct or the Command method. Not only have we created commands, but we've also passed file descriptors to them to receive information back. We learned about the different ways to run a command using the Run or Start method and the multiple ways of retrieving data from the standard output, standard error types, and other file descriptors.

In this chapter, we also discussed the net/http and net/url packages, which are important to be comfortable with when creating HTTP requests to external API servers. Several examples taught us how to create requests with the methods on http.Client, including Do, Get, Post, and PostForm.

It's important to learn how to build robust code, and handling errors gracefully is part of the process. We need to know how to capture errors first, so we discussed how to detect some common errors that can occur when running an external process or sending a request to an external API server. Capturing and handling other errors gives us confidence that our code is ready to take appropriate action when they occur. Finally, we now know how to check for different status codes when the response is not okay.

With all the information learned in this chapter, we should now be more confident in building a CLI that interacts with external commands or sends requests to external APIs. In the next chapter, we'll learn how to write code that can run on multiple different architectures and operating systems.

Questions

1. What method in the `time` package do we use to receive the time after a particular duration via a channel?

2. What is the error type returned from `http.Client`'s Do method?

3. When an HTTP request receives a response with a status code other than `StatusOK`, is the error returned from the request populated?

Answers

1. `time.After(d Duration) <-chan Time`

2. `*url.Error`

3. No

Further reading

- Visit the online documentation for `net/http` at `https://pkg.go.dev/net/http`, and for net/url at `https://pkg.go.dev/net/url`

7

Developing
for Different Platforms

One of the main reasons Go is such a powerful language for building a command-line application is how easy it is to develop an application that can be run on multiple machines. Go provides several packages that allow developers to write code that interacts with the computer independent of the specific operating system. These packages include os, time, path, and runtime. In the first section, we will discuss some commonly used functions in each of these packages and then provide some simple examples to pair with the explanations.

To further drill down the importance of these files, we will revisit the audiofile code and implement a couple of new features that utilize some of the methods that exist in these packages. After all, the best way to learn is by implementing new features with the new functions and methods you've learned about.

We will then learn how to use the runtime library to check the operating system the application is running on and then use that to switch between codes. By learning about build tags, what they are, and how to use them, we will learn about a cleaner way to switch between code blocks to implement a new feature that can be run on three different operating systems: Darwin, Windows, and Linux. By the end of the chapter, you'll feel more confident when building your application, knowing that the code you are writing will work seamlessly, independent of the platform.

In this chapter, we will cover the following key topics:

- Packages for platform-independent functionality
- Implementing independent or platform-specific code
- Build tags for targeted platforms

Technical requirements

- The code files for this chapter are available here: `https://github.com/PacktPublishing/Building-Modern-CLI-Applications-in-Go/tree/main/Chapter07`.

Packages for platform-independent functionality

When you are building a **command-line interface** (**CLI**) that will be shared with the public, it's important that the code is platform-independent to support users who are running the CLI on different operating systems. Golang has supportive packages that provide platform-independent interfaces to operating system functionality. A few of these packages include `os`, `time`, and `path`. Another useful package is the `runtime` package, which helps when detecting the operating system the application is running on, among other things. We will review each of these packages with some simple examples to show how to apply some of the available methods.

The os package

The **operating system** (**os**) package has a Unix-like design but applies uniformly across all operating systems. Think of all the operating system commands you can run in a shell, including external commands. The `os` package is your go-to package. We discussed calling external commands in the previous chapter; now we will discuss this at a higher level and focus on the commands in certain groups: environmental, file, and process operations.

Environmental operations

As the name suggests, the `os` package contains functions that give us information about the environment in which the application is running, as well as change the environment for future method calls. These common operations are for the following working directories:

- `func Chdir(dir string) error`: This changes the current working directory
- `func Getwd() (dir string, err error)`: This gets the current working directory

There are also operations for the environment, as follows:

- `func Environ() []string`: This lists environment keys and values
- `func Getenv(key string) string`: This gets environment variables by key
- `func Setenv(key, value string) error`: This sets environment variables by key and value
- `func Unsetenv(key string) error`: This unsets an environment variable by key
- `func Clearenv()`: This clears environment variables

- `func ExpandEnv(s string)  string`: This expands values of environment variable keys in strings to their values

The *Chapter 7* code exists on GitHub in the `environment.go` file, where we have provided some sample code demonstrating using these operations:

```
func environment() {
    dir, err := os.Getwd()
    if err != nil {
        fmt.Println("error getting working directory:", err)
    }
    fmt.Println("retrieved working directory: ", dir)

    fmt.Println("setting WORKING_DIR to", dir)
    err = os.Setenv("WORKING_DIR", dir)
    if err != nil {
        fmt.Println("error setting working directory:", err)
    }

    fmt.Println(os.ExpandEnv("WORKING_DIR=${WORKING_DIR}"))

    fmt.Println("unsetting WORKING_DIR")
    err = os.Unsetenv("WORKING_DIR")
    if err != nil {
        fmt.Println("error unsetting working directory:", err)
    }

    fmt.Println(os.ExpandEnv("WORKING_DIR=${WORKING_DIR}"))
    fmt.Printf("There are %d environment variables:\n", len(os.
        Environ()))
    for _, envar := range os.Environ() {
        fmt.Println("\t", envar)
    }
}
```

To briefly describe the preceding code, we first get the working directory, then set it to the WORKING_DIR environment variable. To show the change, we utilize `os.ExpandEnv` to print the key-value pair. We then unset the WORKING_DIR environment variable. Again, we show it is unset by using `os.ExpandEnv` to print out the key-value pair. The `os.ExpandEnv` variable will print an empty string if the environment variable is unset. Finally, we print out the count of the environment variables and then range through all to print them. Running the preceding code will produce the following output:

```
retrieved working directory:  /Users/mmontagnino/Code/src/
github.com/marianina8/Chapter-7
setting WORKING_DIR to /Users/mmontagnino/Code/src/github.com/
marianina8/Chapter-7
WORKING_DIR=/Users/mmontagnino/Code/src/github.com/marianina8/
Chapter-7
There are 44 environment variables.
key=WORKING_DIR, value=/Users/mmontagnino/Code/src/github.com/
marianina8/Chapter-7
unsetting WORKING_DIR
WORKING_DIR=
```

If you run this code on your machine rather than Linux, Unix, or Windows, the resulting output will be similar. Try for yourself.

> **Notes on running the following examples**
>
> To run the *Chapter 7* examples, you'll first need to run the install command to install the sleep command to your GOPATH. On Unix-like systems, run the `make install` command followed by the `make run` command. On Linux systems, run the `./build-linux.sh` script followed by the `./run-linux.sh` script. On Windows, run `.\build-windows.ps1` followed by the `.\run-windows.ps1` Powershell script.

File operations

The `os` package also offers a wide variety of file operations that can be applied universally across different operating systems. Many functions and methods can be applied to files, so rather than going over each by name, I will group the functionality and name a few of each:

- The following can be used to change file, directory, and link permissions and owners:
  - `func Chmod(name string, mode FileMode) error`
  - `func Chown(name string uid, gid int) error`
  - `func Lchown(name string uid, gid int) error`

- The following can be used to create pipes, files, directories, and links:

  - `func Pipe() (r *File, w *File, err error)`
  - `func Create(name string) (*File, error)`
  - `func Mkdir(name string, perm FileMode) error`
  - `func Link(oldname, newname string) error`

- The following are used to read from files, directories, and links:

  - `func ReadFile(name string) ([]byte, error)`
  - `func ReadDir(name string) ([]DirEntry, error)`
  - `func Readlink(name string) (string, error)`

- The following retrieve user-specific data:

  - `func UserCacheDir() (string, error)`
  - `func UserConfigDir() (string, error)`
  - func UserHomeDir() (string, error)

- The following are used to write to files:

  - func (f *File) Write(b []byte) (n int, err error)
  - func (f *File) WriteString(s string) (n int, err error)
  - `func WriteFile(name string, data []byte, perm FileMode) error`

- The following are used for file comparison:

  - `func SameFile(fi1, fi2 FileInfo) bool`

There is a `file.go` file within the *Chapter 7* code on GitHub in which we have some sample code using these operations. Within the file are multiple functions, the first, `func createFiles()` `error`, handles the creation of three files to play around with:

```
func createFiles() error {
    filename1 := "file1"
    filename2 := "file2"
    filename3 := "file3"
    f1, err := os.Create(filename1)
    if err != nil {
```

```
        return fmt.Errorf("error creating %s: %v\n", filename1,
            err)
    }
    defer f1.Close()
    f1.WriteString("abc")
    f2, err := os.Create(filename2)
    if err != nil {
        return fmt.Errorf("error creating %s: %v\n", filename2,
            err)
    }
    defer f2.Close()
    f2.WriteString("123")
    f3, err := os.Create(filename3)
    if err != nil {
        return fmt.Errorf("error creating %s: %v", filename3,
            err)
    }
    defer f3.Close()
    f3.WriteString("xyz")
    return nil
}
```

The `os.Create` method allows file creation to work seamlessly on different operating systems. The next function, `file()`, utilizes these files to show how to use methods that exist within the `os` package. The `file()` function primarily gets or changes the current working directory and runs different functions, including the following:

- `func createExamplesDir() (string, error)`: This creates an `examples` directory in the user's home directory

- `func printFiles(dir string) error`: This prints the files/directories under the directory represented by `dir string`

- `func sameFileCheck(f1, f2 string) error`: This checks whether two files, represented by the `f1` and `f2` strings are the same file

Let's first show the `file()` function to get the overall gist of what is going on:

```
originalWorkingDir, err := os.Getwd()
if err != nil {
    fmt.Println("getting working directory: ", err)
```

```go
}
fmt.Println("working directory: ", originalWorkingDir)
examplesDir, err := createExamplesDir()
if err != nil {
    fmt.Println("creating examples directory: ", err)
}
err = os.Chdir(examplesDir)
if err != nil {
    fmt.Println("changing directory error:", err)
}
fmt.Println("changed working directory: ", examplesDir)
workingDir, err := os.Getwd()
if err != nil {
    fmt.Println("getting working directory: ", err)
}
fmt.Println("working directory: ", workingDir)
createFiles()
err = printFiles(workingDir)
if err != nil {
    fmt.Printf("Error printing files in %s\n", workingDir)
}
err = os.Chdir(originalWorkingDir)
if err != nil {
    fmt.Println("changing directory error: ", err)
}
fmt.Println("working directory: ", workingDir)
symlink := filepath.Join(originalWorkingDir, "examplesLink")
err = os.Symlink(examplesDir, symlink)
if err != nil {
    fmt.Println("error creating symlink: ", err)
}
fmt.Printf("created symlink, %s, to %s\n", symlink,
examplesDir)
err = printFiles(symlink)
if err != nil {
    fmt.Printf("Error printing files in %s\n", workingDir)
```

```
    }
    file := filepath.Join(examplesDir, "file1")
    linkedFile := filepath.Join(symlink, "file1")
    err = sameFileCheck(file, linkedFile)
    if err != nil {
        fmt.Println("unable to do same file check: ", err)
    }
    // cleanup
    err = os.Remove(symlink)
    if err != nil {
        fmt.Println("removing symlink error: ", err)
    }
    err = os.RemoveAll(examplesDir)
    if err != nil {
        fmt.Println("removing directory error: ", err)
    }
```

Let's walk through the preceding code. First, we get the current working directory and print it out. Then, we call the createExamplesDir() function and change direction into it.

We then get the current working directory after we change it to ensure it's now the examplesDir value. Next, we call the createFiles() function to create those three files inside the examplesDir folder and call the printFiles() function to list the files in the examplesDir working directory.

We change the working directory back to the original working directory and create a symlink to the examplesDir folder under the home directory. We print the files existing under the symlink to see that they are equal.

Next, we take file0 from examplesDir and file0 from symlink and compare them within the sameFileCheck function to ensure they are equal.

Finally, we run some cleanup functions to remove the symlink and examplesDir folders.

The file function utilizes many methods available in the os package, from getting the working directory to changing it, creating a symlink, and removing files and directories. Showing the separate function call code will give more uses of the os package. First, let's show the code for createExamplesDir:

```
func createExamplesDir() (string, error) {
    homeDir, err := os.UserHomeDir()
    if err != nil {
        return "", fmt.Errorf("getting user's home directory:
         %v\n", err)
```

```go
    }
    fmt.Println("home directory: ", homeDir)
    examplesDir := filepath.Join(homeDir, "examples")
    err = os.Mkdir(examplesDir, os.FileMode(int(0777)))
    if err != nil {
        return "", fmt.Errorf("making directory error: %v\n",
            err)
    }
    fmt.Println("created: ", examplesDir)
    return examplesDir, nil
}
```

The preceding code uses the os package when getting the user's home directory with the os.UserHomeDir method and then creates a new folder with the os.Mkdir method. The next function, printFiles, gets the files to print from the os.ReadDir method:

```go
func printFiles(dir string) error {
    files, err := os.ReadDir(dir)
    if err != nil {
        return fmt.Errorf("read directory error: %s\n", err)
    }
    fmt.Printf("files in %s:\n", dir)
    for i, file := range files {
        fmt.Printf(" %v %v\n", i, file.Name())
    }
    return nil
}
```

Lastly, sameFileCheck takes two files represented by strings, f1 and f2. To get the file info for each file, the os.Lstat method is called on the file string. os.SameFile takes this file info and returns a boolean value to symbolize the result – true if the files are the same and false if not:

```go
func sameFileCheck(f1, f2 string) error {
    fileInfo0, err := os.Lstat(f1)
    if err != nil {
        return fmt.Errorf("getting fileinfo: %v", err)
    }
    fileInfo0Linked, err := os.Lstat(f2)
    if err != nil {
```

```
        return fmt.Errorf("getting fileinfo: %v", err)
    }
    isSameFile := os.SameFile(fileInfo0, fileInfo0Linked)
    if isSameFile {
        fmt.Printf("%s and %s are the same file.\n", fileInfo0.
            Name(), fileInfo0Linked.Name())
    } else {
    fmt.Printf("%s and %s are NOT the same file.\n", fileInfo0.
        Name(), fileInfo0Linked.Name())
    }
    return nil
}
```

This concludes the code samples utilizing methods from the `os` package related to file operations. Next, we will discuss some operations related to processes running on the machine.

Process operations

When calling external commands, we can get a **process ID (pid)**, associated with the process. Within the `os` package, we can perform actions on the process, send the process signals, or wait for the process to complete and then receive a process state with information regarding the process that was completed. In the *Chapter 7* code, we have a `process()` function, which utilizes some of the following methods for processes and process states:

- `func Getegid() int`: This returns the effective group ID of the caller. Note, this is not supported in Windows, the concept of group IDs is specific to Unix-like or Linux systems. For example, this will return −1 on Windows.

- `func Geteuid() int`: This returns the effective user ID of the caller. Note, this is not supported in Windows, the concept of user IDs is specific to Unix-like or Linux systems. For example, this will return -1 on Windows.

- `func Getpid() int`: This gets the process ID of the caller.

- `func FindProcess(pid int) (*Process, error)`: This returns the process associated with the `pid`.

- `func (p *Process) Wait() (*ProcessState, error)`: This returns the process state when the process completes.

- `func (p *ProcessState) Exited() bool`: This returns `true` if the process exited.

- `func (p *ProcessState) Success() bool`: This returns `true` if the process exited successfully.

- `func (p *ProcessState) ExitCode() int`: This returns the exit code of the process.

- `func (p *ProcessState) String() string`: This returns the process state in string format.

The code is as follows and starts with several print line statements that return the caller's effective group, user, and process ID. Next, a cmd sleep command is defined. The command is started and from the cmd value, we get the pid:

```
func process() {
    fmt.Println("Caller group id:", os.Getegid())
    fmt.Println("Caller user id:", os.Geteuid())
    fmt.Println("Process id of caller", os.Getpid())
    cmd := exec.Command(filepath.Join(os.Getenv("GOPATH"),
            "bin", "sleep"))
    fmt.Println("running sleep for 1 second...")
    if err := cmd.Start(); err != nil {
        panic(err)
    }
    fmt.Println("Process id of sleep", cmd.Process.Pid)
    this, err := os.FindProcess(cmd.Process.Pid)
    if err != nil {
        fmt.Println("unable to find process with id: ", cmd.
            Process.Pid)
    }
    processState, err := this.Wait()
    if err != nil {
        panic(err)
    }
    if processState.Exited() && processState.Success() {
        fmt.Println("Sleep process ran successfully with exit
            code: ", processState.ExitCode())
    } else {
        fmt.Println("Sleep process failed with exit code: ",
            processState.ExitCode())
    }
    fmt.Println(processState.String())
}
```

From the process' pid, we then can find the process using the os.FindProcess method. We call the Wait() method in the process to get os.ProcessState. This Wait() method, like the cmd.Wait() method, waits for the process to complete. Once completed, the process state is returned. We can check whether the process state is exited with the Exited() method and whether it was successful with the Success() method. If so, we print that the process ran successfully along with the exit code, which we get from the ExitCode() method. Finally, the process state can be printed cleanly with the String() method.

The time package

Operating systems provide access to time via two different types of internal clocks:

- **A wall clock**: This is used for telling the time and is subject to variations due to clock synchronization with the **Network Time Protocol (NTP)**

- **A monotonic clock**: This is used for measuring time and is not subject to variations due to clock synchronization

To be more specific on the variations, if the wall clock notices that it is moving faster or slower than the NTP, it will adjust its clock rate. The monotonic clock will not adjust. When measuring durations, it's important to use the monotonic clock. Luckily with Go, the Time struct contains both the wall and monotonic clocks, and we don't need to specify which is used. Within the *Chapter 7* code, there is a timer.go file, which shows how to get the current time and duration, regardless of the operating system:

```
func timer() {
    start := time.Now()
    fmt.Println("start time: ", start)
    time.Sleep(1 * time.Second)
    elapsed := time.Until(start)
    fmt.Println("elapsed time: ", elapsed)
}
```

When running the following code, you'll see a similar output:

```
start time:   2022-09-24 23:47:38.964133 -0700 PDT
m=+0.000657043
elapsed time:   -1.002107875s
```

Also, many of you have also seen that there is a time.Now().Unix() method. It returns to the epoch time, or time that has elapsed since the Unix epoch, January 1, 1970, UTC. These methods will work similarly regardless of the operating system and architecture they are run on.

The path package

When developing a command-line application for different operating systems, you'll most likely have to deal with handling file or directory path names. In order to handle these appropriately across different operating systems, you'll need to use the path package. Because this package does not handle Windows paths with drive letters or backslashes, as we used in the previous examples, we'll use the path/filepath package.

The path/filepath package uses either forward or back slashes depending on the operating system. Just for fun, within the *Chapter 7* walking.go file, I've used the filepath package to walk through a directory. Let's look at the code:

```go
func walking() {
    workingDir, err := os.Getwd()
    if err != nil {
        panic(err)
    }
    dir1 := filepath.Join(workingDir, "dir1")
    filepath.WalkDir(dir1, func(path string, d fs.DirEntry, err
      error) error {
        if !d.IsDir() {
            contents, err := os.ReadFile(path)
            if err != nil {
                return err
            }
            fmt.Printf("%s -> %s\n", d.Name(),
                string(contents))
        }
        return nil
    })
}
```

We get the current working directory with os.Getwd(). Then create a path for the dir1 directory that can be used for any operating system using the filepath.Join method. Finally, we walk the directory using filepath.WalkDir and print out the filename and its contents.

The runtime package

The final package to discuss within this section is the runtime package. It's mentioned because it's used to easily determine the operating system the code is running on and therefore execute blocks of code, but there's so much information you can get from the runtime system:

- GOOS: This returns the running application's operating system target
- GOARCH: This returns the running application's architecture target
- func GOROOT() string: This returns the root of the Go tree
- Compiler: This returns the name of the compiler toolchain that built the binary
- func NumCPU() int: This returns the number of logical CPUs usable by the current process
- func NumGoroutine() int: This returns the number of goroutines that currently exist
- func Version() string: This returns the Go tree's version string

This package will provide you with enough information to understand the runtime environment. Within the *Chapter 7* code in the checkRuntime.go file is the checkRuntime function, which puts each of these into practice:

```go
func checkRuntime() {
    fmt.Println("Operating System:", runtime.GOOS)
    fmt.Println("Architecture:", runtime.GOARCH)
    fmt.Println("Go Root:", runtime.GOROOT())
    fmt.Println("Compiler:", runtime.Compiler)
    fmt.Println("No. of CPU:", runtime.NumCPU())
    fmt.Println("No. of Goroutines:", runtime.NumGoroutine())
    fmt.Println("Version:", runtime.Version())
    debug.PrintStack()
}
```

Running the code will provide a similar output to the following:

```
Operating System: darwin
Architecture: amd64
Go Root: /usr/local/go
Compiler: gc
No. of CPU: 10
No. of Goroutines: 1
Version: go1.19
```

```
goroutine 1 [running]:
runtime/debug.Stack()
        /usr/local/go/src/runtime/debug/stack.go:24 +0x65
runtime/debug.PrintStack()
        /usr/local/go/src/runtime/debug/stack.go:16 +0x19
main.checkRuntime()
        /Users/mmontagnino/Code/src/github.com/marianina8/
Chapter-7/checkRuntime.go:17 +0x372
main.main()
        /Users/mmontagnino/Code/src/github.com/marianina8/
Chapter-7/main.go:9 +0x34
```

Now that we have learned about some of the packages required for building a command-line application that runs across multiple operating systems and architectures, in the next section, we'll return to the `audiofile` CLI from previous chapters and implement a few new functions and show how the methods and functions we've learned in this section can come into play.

Implementing independent or platform-specific code

The best way to learn is to put what has been learned into practice. In this section, we'll revisit the `audiofile` CLI to implement a few new commands. In the code for the new features we'll implement, the focus will be on the use of the `os` and `path/filepath` packages.

Platform-independent code

Let's now implement a few new features for the `audiofile` CLI that will run independently of the operating system:

- `Delete`: This deletes stored metadata by ID
- `Search`: This searches stored metadata for a specific search string

The creation of each of these new feature commands was initiated with the cobra-CLI; however, the platform-specific code is isolated in the `storage/flatfile.go` file, which is the flat file storage for the storage interface.

First, let's show the `Delete` method:

```
func (f FlatFile) Delete(id string) error {
    dirname, err := os.UserHomeDir()
    if err != nil {
        return err
```

```
    }
    audioIDFilePath := filepath.Join(dirname, "audiofile", id)
    err = os.RemoveAll(audioIDFilePath)
    if err != nil {
        return err
    }
    return nil
}
```

The flat file storage is stored under the user's home directory under the `audiofile` directory. Then, as each new audio file and matching metadata is added, it is stored within its unique identifier ID. From the os package, we use `os.UserHomeDir()` to get the user's home directory and then use the `filepath.Join` method to create the required path to delete all the metadata and files associated with the ID independent of the operating system. Make sure you have some audiofiles stored locally in the flat file storage. If not, add a few files. For example, use the `audio/beatdoctor.mp3` file and upload using the following command:

```
./bin/audiofile upload --filename audio/beatdoctor.mp3
```

The ID is returned after a successful upload:

```
Uploading audio/beatdoctor.mp3 ...
Audiofile ID:   a5d9ab11-6f5f-4da0-9307-a3b609b0a6ba
```

You can ensure that the data has been added by running the `list` command:

```
./bin/audiofile list
```

The `audiofile` metadata is returned, so we have double-checked its existence in storage:

```
    {
        "Id": "a5d9ab11-6f5f-4da0-9307-a3b609b0a6ba",
        "Path": "/Users/mmontagnino/audiofile/a5d9ab11-6f5f-
4da0-9307-a3b609b0a6ba/beatdoctor.mp3",
        "Metadata": {
            "tags": {
                "title": "Shot In The Dark",
                "album": "Best Bytes Volume 4",
                "artist": "Beat Doctor",
                "album_artist": "Toucan Music (Various
Artists)",
```

```
                    "composer": "",
                    "genre": "Electro House",
                    "year": 0,
                    "lyrics": "",
                    "comment": "URL: http://freemusicarchive.org/
music/Beat_Doctor/Best_Bytes_Volume_4/09_beat_doctor_shot_in_
the_dark\r\nComments: http://freemusicarchive.org/\r\nCurator:
Toucan Music\r\nCopyright: Attribution-NonCommercial 3.0
International: http://creativecommons.org/licenses/by-nc/3.0/"
                    },
                    "transcript": ""
            },
            "Status": "Complete",
            "Error": null
        },
```

Now, we can delete it:

```
./bin/audiofile delete --id a5d9ab11-6f5f-4da0-9307-
a3b609b0a6ba
success
```

Then confirm that it's been deleted by trying to get the audio by ID:

```
./bin/audiofile get --id a5d9ab11-6f5f-4da0-9307-a3b609b0a6ba
Error: unexpected response: 500 Internal Server Error
Usage:
  audiofile get [flags]

Flags:
  -h, --help        help for get
      --id string   audiofile id

unexpected response: 500 Internal Server Error%
```

Looks like an unexpected error has occurred, and we haven't properly implemented how to handle this when searching for metadata for a file that has been deleted. We'll need to modify the `services/metadata/handler_getbyid.go` file. At line 20, where we call the `GetById` method and handle the error, let's return `200` instead of `500` after confirming the error is related to a folder not being found. It's not necessarily an error that the user is searching for an ID that does not exist:

```
audio, err := m.Storage.GetByID(id)
if err != nil {
    if strings.Contains(err.Error(), "not found")
||     strings.Contains(err.Error(), "no such file or
directory") {
        io.WriteString(res, "id not found")
        res.WriteHeader(200)
        return
    }
    res.WriteHeader(500)
    return
}
```

Let's try it again:

./bin/audiofile get --id a5d9ab11-6f5f-4da0-9307-a3b609b0a6ba
id not found

That's much better! Now let's implement the search functionality. The implementation again is isolated to the `storage/flatfile.go` file where you will find the `Search` method:

```
func (f FlatFile) Search(searchFor string) ([]*models.Audio,
error) {
    dirname, err := os.UserHomeDir()
    if err != nil {
        return nil, err
    }
    audioFilePath := filepath.Join(dirname, "audiofile")
    matchingAudio := []*models.Audio{}
    err = filepath.WalkDir(audioFilePath, func(path string,
            d fs.DirEntry, err error) error {
        if d.Name() == "metadata.json" {
            contents, err := os.ReadFile(path)
            if err != nil {
                return err
```

```
        }
        if strings.Contains(strings.
            ToLower(string(contents)), strings.
            ToLower(searchFor)) {
              data := models.Audio{}
              err = json.Unmarshal(contents, &data)
              if err != nil {
                  return err
              }
              matchingAudio = append(matchingAudio, &data)
        }
      }
      return nil
    })
    return matchingAudio, err
}
```

Like most of the methods existing in the storage, we start by getting the user's home directory with the `os.UserHomeDir()` method and then, again, use `filepath.Join` to get the root `audiofile` path directory, which we will be walking. The `filepath.WalkDir` method is called starting at `audioFilePath`. We check each of the `metadata.json` files to see whether the `searchFor` string exists within the contents. The method returns a slice of `*models.Audio` and if the `searchFor` string is found within the contents, the audio is appended onto the slice that will be returned later.

Let's give this a try with the following command and see that the expected metadata is returned:

```
./bin/audiofile search --value "Beat Doctor"
```

Now that we've created a few new commands to show how the `os` package and `path/filepath` packages can be used in a real-life example, let's try to write some code that can run specifically on one operating system or another.

Platform-specific code

Suppose your command-line application requires an external application that exists on the operating system, but the application required differs between operating systems. For the `audiofile` command-line application, suppose we want to create a command to play the audio file via the command line. Each operating system will need to use a different command to play the audio, as follows:

- macOS: `afplay <filepath>`
- Windows: `start <filepath>`
- Linux: `aplay <filepath>`

Again, we use the Cobra-CLI to create the new `play` command. Let's look at each different function that would need to be called for each operating system to play the audio file. First is the code for macOS:

```go
func darwinPlay(audiofilePath string) {
    cmd := exec.Command("afplay", audiofilePath)
    if err := cmd.Start(); err != nil {
        panic(err)
    }
    fmt.Println("enjoy the music!")
    err := cmd.Wait()
    if err != nil {
        panic(err)
    }
}
```

We create a command to use the `afplay` executable and pass in the `audiofilePath`. Next is the code for Windows:

```go
func windowsPlay(audiofilePath string) {
    cmd := exec.Command("cmd", "/C", "start", audiofilePath)
    if err := cmd.Start(); err != nil {
        return err
    }
    fmt.Println("enjoy the music!")
    err := cmd.Wait()
    if err != nil {
        return err
    }
}
```

This is a very similar function, except it uses the `start` executable in Windows to play the audio. Last is the code for Linux:

```go
func linuxPlay(audiofilePath string) {
    cmd := exec.Command("aplay", audiofilePath)
    if err := cmd.Start(); err != nil {
        panic(err)
    }
    fmt.Println("enjoy the music!")
```

```
        err := cmd.Wait()
        if err != nil {
            panic(err)
        }
    }
```

Again, the code is practically identical except for the application which is called to play the audio. In another case, this code could be more specific for the operating system, require different arguments, and even require a full path specific to the operating system. Regardless, we are ready to use these functions within the `play` command's `RunE` field. The full `play` command is as follows:

```
var playCmd = &cobra.Command{
    Use: "play",
    Short: "Play audio file by id",
    RunE: func(cmd *cobra.Command, args []string) error {
        b, err := getAudioByID(cmd)
        if err != nil {
            return err
        }
        audio := models.Audio{}
        err = json.Unmarshal(b, &audio)
        if err != nil {
            return err
        }
        switch runtime.GOOS {
        case "darwin":
            darwinPlay(audio.Path)
            return nil
        case "windows":
            windowsPlay(audio.Path)
            return nil
        case "linux":
            linuxPlay(audio.Path)
            return nil
        default:
            fmt.Println(`Your operating system isn't supported
                for playing music yet.
                Feel free to implement your additional use
                case!`)
```

```
        }
        return nil
    },
}
```

The important part of this code is that we have created a switch case for the `runtime.GOOS` value, which tells us what operating system the application is running on. Depending on the operating system, a different method is called to start a process to play the audio file. Let's recompile and try the play method with one of the stored audio file IDs:

```
./bin/audiofile play --id bf22c5c4-9761-4b47-aab0-47e93d1114c8
enjoy the music!
```

The final section of this chapter will show us how to implement this differently, if we'd like to, using build tags.

Build tags for targeted platforms

Built tags, or build constraints, can be used for many purposes, but in this section, we will be discussing how to use build tags to identify which files should be included in a package when building for specific operating systems. Build tags are given in a comment at the top of a file:

```
//go:build
```

Build tags are passed in as flags when running `go build`. There could be more than one tag on a file, and they follow on from the comment with the following syntax:

```
//go:build [tags]
```

Each tag is separated by a space. Suppose we want to indicate that this file will only be included in a build for the Darwin operating system, then we would add this to the top of the file:

```
//go:build darwin
```

Then when building the application, we would use something like this:

```
go build -tags darwin
```

This is just a super quick overview of how build tags can be used to constrain files specific to operating systems. Before we go into an implementation of this, let's discuss the `build` package in a bit more detail.

The build package

The `build` package gathers information about Go packages. In the *Chapter07* code repository, there is a `buildChecks.go` file, which uses the `build` package to get information about the current package. Let's see what information this code can give us:

```
func buildChecks() {
    ctx := build.Context{}
    p1, err := ctx.Import(".", ".", build.AllowBinary)
    if err != nil {
        fmt.Println("err: ", err)
    }
    fmt.Println("Dir:", p1.Dir)
    fmt.Println("Package name: ", p1.Name)
    fmt.Println("AllTags: ", p1.AllTags)
    fmt.Println("GoFiles: ", p1.GoFiles)
    fmt.Println("Imports: ", p1.Imports)
    fmt.Println("isCommand: ", p1.IsCommand())
    fmt.Println("IsLocalImport: ", build.IsLocalImport("."))
    fmt.Println(ctx)
}
```

We first create the `context` variable and then call the `Import` method. The `Import` method is defined in the documentation as follows:

```
func (ctxt *Context) Import(path string, srcDir string, mode
ImportMode) (*Package, error)
```

It returns the details about the Go package named by the `path` and `srcDir` source directory parameters. In this case, the `main` package is returned from the package, then we can check all the variables and methods that exist to get more information on the package. Running this method locally will return something like this:

```
Dir: .
Package name:  main
AllTags:  [buildChecks]
```

```
GoFiles:   [checkRuntime.go environment.go file.go main.go
process.go timer.go walking.go]
Imports:   [fmt io/fs os os/exec path/filepath runtime runtime/
debug strings time]
isCommand/main package:   true
IsLocalImport:   true
```

Most of the values we are checking are self-explanatory. `AllTags` returns all tags that exist within the `main` package. `GoFiles` returns all the files included in the `main` package. `Imports` are all the unique imports that exist within the package. `IsCommand()` returns `true` if the package is considered a command to be installed, or if it is the main package. Finally, the `IsLocalImport` method checks whether an import file is local. This is a fun extra detail to interest you more about what the `build` package could potentially offer you.

Build tags

Now that we have learned a little bit more about the `build` package, let's use it for the main purpose of this chapter, building packages for specific operating systems. Build tags should be named intentionally, and since we are using them for a specific purpose, we can name each build tag by an operating system:

```
//go:build darwin
//go:build linux
//go:build windows
```

Let's revisit the audio file code. Remember how in the `play` command, we check the `runtime` operating system and then call a specific method. Let's rewrite this code using build tags.

Example in the audio file

Let's first simplify the command's code to the following:

```
var playCmd = &cobra.Command{
    Use: "play",
    Short: "Play audio file by id",
    Long: `Play audio file by id`,
    RunE: func(cmd *cobra.Command, args []string) error {
        b, err := getAudioByID(cmd)
        if err != nil {
            return err
        }
        audio := models.Audio{}
```

```
        err = json.Unmarshal(b, &audio)
        if err != nil {
            return err
        }
        return play(audio.Path)
    },
}
```

Basically, we've simplified the code greatly by removing the operating system switch statement and the three functions that implement the play feature for each operating system. Instead, we've taken the code and created three new files: play_darwin.go, play_windows.go, and play_linux.go. Within each of these files is a build tag for each operating system. Let's take the Darwin file, play_darwin.go, for example:

```
//go:build darwin

package cmd

import (
    "fmt"
    "os/exec"
)
func play(audiofilePath string) error {
    cmd := exec.Command("afplay", audiofilePath)
    if err := cmd.Start(); err != nil {
        return err
    }
    fmt.Println("enjoy the music!")
    err := cmd.Wait()
    if err != nil {
        return err
    }
    return nil
}
```

Notice that the play function has been renamed to match the function called in the play command in play.go. Since only one of the files gets included in the build, there's no confusion as to which play function is called. We ensure that only one gets called within the make file, which is how we are currently running the application. In Makefile, I've designated a command to build specifically for Darwin:

```
build-darwin:
    go build -tags darwin -o bin/audiofile main.go
    chmod +x bin/audiofile
```

A Go file containing the play function is created for Windows and Linux. The specific tags for each operating system will similarly need to be passed into the -tags flag when building your application. In later chapters, we will discuss cross-compiling, which is the next step. But before we do, let's leave this chapter by reviewing a list of OS-level differences to keep in mind while developing for multiple platforms.

OS-level differences

Since you'll be building your application for the main operating systems, it's important to know the differences between them and know what to look out for. Let's dive in with the following list:

- **Filesystem**:

  - Windows uses a different filesystem than Linux and Unix, so be mindful of the file paths when accessing files in your Go code.

  - File paths in Windows use backslashes, (\), as directory separators, while Linux and Unix use forward slashes (/).

- **Permissions**:

  - Unix-like systems use file modes to manage permissions, where permissions are assigned to files and directories.

  - Windows uses an **access control list** (ACL) to manage permissions, where permissions are assigned to specific users or groups for a file or directory in a more flexible and granular manner.

  - In general, it's a good practice to carefully consider user and group permissions when developing any command-line application, regardless of the operating system it will be running on.

- **Executing commands**:

  - The exec package in Go provides a convenient way to run commands in the same manner as in the terminal. However, it's important to note that the command and its arguments must be passed in the correct format for each operating system.

  - On Windows, you need to specify the file extension (for example, .exe, .bat, etc.) to run an executable file.

- **Environmental variables**:

  - Environmental variables can be used to configure your application, but their names and values may be different between Windows and Linux/Unix.

  - On Windows, environmental variable names are case-insensitive, while on Linux/Unix, they are case-sensitive.

- **Line endings**:

  - Windows uses a different line ending character than Linux/Unix, so be careful when reading or writing files in your Go code. Windows uses a carriage return (`\r`) followed by a line feed (`\n`), while Linux/Unix uses only a line feed (`\n`).

- **Signal handling**:

  - In Unix systems, the `os/signal` package provides a way to handle signals sent to your application. However, this package is not supported on Windows.

  - To handle signals in a cross-platform way, you can use the `os/exec` package instead.

- **User input**:

  - The way user input is read may also be different between Windows and Linux/Unix. On Windows, you may need to use the `os.Stdin` property, while on Linux/Unix you can use `os.Stdin` or the `bufio` package to read user input.

- **Console colors**:

  - On Windows, the console does not support ANSI escape codes for changing text color, so you will need to use a different approach for coloring text in the console.

  - There are libraries available in Go, such as `go-colorable`, that provide a platform-independent way to handle console colors.

- **Standard streams**:

  - Standard streams, such as `os.Stdin`, `os.Stdout`, and `os.Stderr` may behave differently between Windows and Linux/Unix. It's important to test your code on both platforms to make sure it works as expected.

These are some of the differences to be aware of when developing a command-line application in Go for different operating systems. It's important to thoroughly test your application on each platform to ensure it behaves as expected.

Summary

The more operating systems your application supports, the more complicated it will get. Hopefully armed with the knowledge of some supportive packages for developing independently of the platform, you'll feel confident that your application will run similarly across different operating systems. Also, by checking the runtime operating system and even separating code into separate operating system-specific files with build tags, you have at least a couple of options for defining how to organize your code. This chapter goes more in-depth than may be necessary, but hopefully, it inspires you.

Building for multiple operating systems will expand the usage of your command-line application. Not only can you reach Linux or Unix users but also Darwin and Windows users as well. If you want to grow your user base, then building an application to support more operating systems is an easy way to do so.

In the next chapter, *Chapter 8, Building for Humans Versus Machines*, we'll learn how to build a CLI that outputs according to who is receiving it: a machine or human. We'll also learn how to structure the language for clarity and name commands for consistency with the rest of the CLIs in the community.

Questions

1. What are the two different clocks that exist within an operating system? And does the time.Time struct in Go store one or the other clock, or both? Which should be used for calculating duration?

2. Which package constant can be used to determine the runtime operating system?

3. Where is the build tag comment set within a Go file – at the top, bottom, or above the defined function?

Answers

1. The wall clock and monotonic clock. The time.Time struct stores both time values. The monotonic clock value should be used when calculating duration.

2. runtime.GOOS

3. At the top first line of the Go file.

Further reading

- Visit the online documentation for the packages discussed at https://pkg.go.dev/.

Part 3: Interactivity and Empathic Driven Design

This part is about how to develop a more user-friendly command-line interface (CLI) by considering the end user's perspective. It covers topics such as building for humans versus machines, using ASCII art to improve information density, and ensuring consistency in flag names and arguments. The section also emphasizes the importance of empathy in CLI development, including rewriting errors in a user-friendly way, providing detailed logging, and creating man pages and usage examples. Additionally, the benefits of interactivity through prompts and terminal dashboards are discussed, with examples of how to build user prompts and dashboards using the Termdash library.

This part has the following chapters:

- *Chapter 8, Building for Humans Versus Machines*
- *Chapter 9, The Empathic Side of Development*
- *Chapter 10, Interactivity with Prompts and Terminal Dashboards*

8

Building for Humans versus Machines

Thinking about your end user while you develop your command-line application will make you a more empathic developer. Consider not just how you feel about the way certain **command-line interfaces** (**CLIs**) behave but also how you could improve the experience for yourself and others. Much goes into usability and it's not possible to cram it all into a single chapter, so we suggest following up with the suggested article and book in the *Further reading* section.

One of the first points to consider when building your command-line interface is that while it will be primarily used by humans, it can also be called within scripts, and the output from your program could be used as input into another application, such as **grep** or **awk**. Within this chapter, we'll go over how to build for both and how to tell when you're outputting to one versus the other.

The second point is the use of ASCII art to increase information density. Whether you're outputting data as a table, or adding color or emojis, the idea is to make information jump out of the terminal in a way that the end user can quickly understand the data presented to them.

Finally, consistency also increases clarity for your users. When your CLI uses consistency within flag names and positional arguments across different commands and subcommands, your user can feel more confident in the steps they need to take when navigating your CLI. By the end of the chapter, you'll hopefully have more to consider when building your CLI and be prompted to make usability improvements. Within this chapter, we'll cover the following topics:

- Building for humans versus machines
- Increasing information density with ASCII art
- Being consistent across CLIs

Technical requirements

You'll need a Unix operating system to understand and run the examples shared in the chapter.

You can also find the code examples on GitHub at `https://github.com/PacktPublishing/Building-Modern-CLI-Applications-in-Go/tree/main/Chapter08`.

Building for humans versus machines

CLIs have a long history where their interactions were tailored for other programs and machines. Their design was more similar to functions within a program than a graphical interface. Because of this, many Unix programs today still operate under the assumption that they will be interacting with another program.

Today, however, CLIs are more often used by humans than other machines while still carrying an outdated interaction design. It's time that we built CLIs for their primary user—the human.

In this section, we will compare the machine-first design to the human-first design and learn how to check whether you are outputting to the TTY. As we can recall from *Chapter 1*, *Understanding CLI Standards*, **TTY** is short for **TeleTYpewriter**, which evolved into the input and output device to interact with large mainframes. In today's world, desktop environments for operating systems, or **OSs** for short, provide a terminal window. This terminal window is a virtual teletypewriter. They are often called **pseudo-teletypes**, or **PSY** for short. It's also an indication that a human is on the other end, versus a program.

Is it a TTY?

First, let's understand devices. **Devices** can be anything from hard drives, RAM disks, DVD players, keyboards, mouses, printers, tape drivers, to TTYs. A **device driver** provides the interface between the operating system and the device; it provides an API that the operating system understands and accepts.

Figure 8.1 – Figure showing communication from OS to the TTY device via a device driver

On Unix-based OSs, there are two major device drivers:

- **Block** – interfaces for devices such as hard drives, RAM disks, and DVD players

- **Character** – interfaces for the keyboard, mouse, printers, tape drivers, TTYs, and so on

If you check that the standard input, **stdin**, or standard output, **stdout**, is a **character** device, then you can assume that you are receiving input from or sending output to a human.

Is it a TTY on a Unix or Linux operating system?

In a terminal, if you type the `tty` command, it will output the file name connected to **stdin**. Effectively, it is the number of the terminal window.

Let's run the command in our Unix terminal window and see what the result is:

```
mmontagnino@Marians-MacBook-Pro marianina8 % tty
/dev/ttys014
```

There is a shorthand silent, `-s`, flag that can be used to suppress output. However, the application still returns an exit code:

- Exit code 0 – standard input is coming from a TTY

- Exit code 1 – standard input is not coming from a TTY

- Exit code 2 – syntax error from invalid parameters

- Exit code 3 – a write error

In Unix, typing `&&` after a command means that the second command will only execute if the first command runs successfully, with exit code 0. So, let's try this code to see if we're running in a TTY:

```
mmontagnino@Marians-MacBook-Pro marianina8 % tty -s && echo
"this is a tty"
this is a tty
```

Since we ran those commands in a terminal, the result is `this is a tty`.

Programmatically check on a Unix or Linux operating system

There are a few ways to do this programmatically. We can use the code located in the `Chapter-8/isatty.go` file:

```go
func IsaTTY() {
    fileInfo, _ := os.Stdout.Stat()
    if (fileInfo.Mode() & os.ModeCharDevice) != 0 {
        fmt.Println("Is a TTY")
    } else {
        fmt.Println("Is not a TTY")
```

```
    }
  }
```

The preceding code grabs the file info from the standard output, **stdout**, file with the following code:

```
fileInfo, _ := os.Stdout.Stat()
```

Then, we check the result of a bitwise operation, &, between `fileInfo.Mode()` and `os.ModeCharDevice`. The bitwise operator, &, copies a bit to the result if it exists in both operands.

Let's take a quite simple example: 7 & 6 within a truth table. 7 values are represented by binary 111 and 6 values are represented by 110.

Input		Output
7	**6**	**&**
1	1	1
1	1	1
1	0	0

Figure 8.2 – Truth table to show the & operation calculation

The & operation checks each bit and whether they are the same, and if so, carry a bit over, or 1. If the bits differ, no bit is carried over, or 0. The resulting value is 110.

Now, in our more complicated example, the following code, `fileInfo.Mode() & os.ModeCharDevice`, performs a bitwise operation between `fileInfo.Mode()` and `os.ModeCharDevice`. Let's look at what this operation looks like when the code standard output is connected to a terminal:

Is a TTY	
Code	**Value**
`fileInfo.Mode()`	Dcrw--w----
`os.ModeCharDevice`	c---------
`fileInfo.Mode() & os.ModeCharDevice`	c---------
`(fileInfo.Mode() & os.ModeCharDevice) != 0`	TRUE

Figure 8.3 – The code next to its value when standard output is connected to a TTY

In *Figure 8.3*, the file mode of the standard output is defined by the `fileInfo.Mode()` method call; its value is **Dcrw--w----**. If you look at the documentation for the **os** package at https://pkg.go.dev/os, you will see that the os.ModeDevice, **D**, bit is set to indicate that the file is a device file, followed by the os.ModeCharDevice, **c**, bit set to indicate that it is a Unix character device. When we do a bitwise operation against the mode of stdin against os.ModCharDevice, we see that the same bits are carried over and the result does not equal zero, hence (fileInfo.Mode() & os.ModeCharDevice) != 0 is **true**, and the device is a TTY.

What would this code look like if the output were piped into another process? Let's look:

Is not a TTY	
Code	Value
fileInfo.Mode()	prw-rw----
os.ModeCharDevice	c---------
fileInfo.Mode() & os.ModeCharDevice	----------
(fileInfo.Mode() & os.ModeCharDevice) != 0	FALSE

Figure 8.4 – The code next to its value when standard output is not connected to a TTY

Now the standard output's value is **prw-rw----**. The os.ModeNamedPipe, **p**, bit is set to indicate that it is connected to a **pipe**, a redirection to another process. When the bitwise operation is performed against os.ModeCharDevice, we see that no bits are copied over, hence (fileInfo.Mode() & os.ModeCharDevice) != 0 is **false**, and the device is not a TTY.

Programmatically check on any operating system

We suggest using a package that has already gone through the trouble of determining the code for a larger set of operating systems to check whether standard output is sent to a TTY. The most popular package we found was github.com/mattn/go-isatty, which we used in the Chapter-8/utils/isatty.go file:

```
package utils

import (
  "fmt"
  "os"
  isatty "github.com/mattn/go-isatty"
)

func IsaTTY() {
  if isatty.IsTerminal(os.Stdout.Fd()) ||  isatty.
    IsCygwinTerminal(os.Stdout.Fd()) {
    fmt.Println("Is a TTY")
  } else {
    fmt.Println("Is not a TTY")
```

```
        }
    }
```

Now that we know whether we are outputting to a TTY, which indicates that there is a human on the other end, versus not a TTY, we can tailor our output accordingly.

Designing for a machine

As aforementioned, CLIs were originally designed for machines first. It is important to understand what it exactly means to design another program. Although we would want to tailor our applications toward a human-first design, there will be times when we would need to output in a way that can easily be passed as input to the `grep` or `awk` command, because other applications will expect streams of either plain or JSON text.

Users will be using your CLI in many unexpected ways. Some of those ways are often within a bash script that pipes the output of your command as input into another application. If your application, as it should, outputs in the human-readable format first, it needs to also output in machine-readable format when the standard input is not connected to a TTY terminal. In the latter case, make sure any color and ASCII art, in the form of progress bars, for example, are disabled. The text should also be single-lined tabular data that can easily be integrated with the `grep` and `awk` tools.

Also, it is important that you offer several persistent flags for your users to output in machine-readable output when necessary:

- `--plain`, for outputting plain text with one record of data per line
- `--json`, for outputting JSON text that can be piped to and from the curl command
- `--quiet`, `-q`, or `--silent`, `-s`, for suppressing nonessential output

Provide plain text when it does not impact usability. In other cases, offer the optional previous flags to give the user the ability to pipe its output easily into the input of another.

Designing for a human

The modern command-line application is designed for its primary consumer—the human. This may seemingly complicate the interface because there's a bit more to consider. The way data is output and how quickly the data is returned can affect how a user perceives the quality and robustness of your CLI. We'll go over some key areas of design:

- Conversation as the norm
- Empathy
- Personalization
- Visual language

Let's go into each in more detail so we can fully understand how this impacts a human-centred design.

Conversation as the norm

Since your CLI will be responding to a human and not another program, interaction should flow like a conversation. As an application leans toward a conversational language, the user will feel more at ease. Consider your application as the guide, as well, toward usage of the CLI.

When a user runs a command and is missing important flags or arguments, then your application can prompt for these values. Prompts, or surveys, are a way to include a conversational back-and-forth flow of asking questions and receiving answers from the user. However, prompts should not be a requirement as flags and arguments should be available options for your commands. We will be going over prompts in more detail in *Chapter 10, Interactivity with Prompts and Terminal Dashboards*.

If your application contains a state, then communicate the current state similar to how `git` provides a `status` command and notifies the user when any commands change the state. Similarly, if your application provides workflows, typically defined by a chain of commands, then you can suggest commands to run next.

Being succinct is important when communicating with your user. Just like in conversation, if we muddle our words with too much extraneous information, people can become confused about the point we are trying to make. By communicating what's important, but keeping it brief, our users will get the most important information quickly.

Context is important. If you are communicating with an end user versus a developer, that makes a difference. In that case, unless you are in verbose mode, there's no reason to output anything only a developer would understand.

If the user is doing anything dangerous, ask for confirmation and match the level of confirmation with the level of danger that can be invoked by the command:

- **Mild**:

  - Example: deleting a file

  - Confirmation:

    - If the command is a `delete` command, you don't need to confirm

    - If not a `delete` command, prompt for confirmation

- **Moderate**:

  - Example: deleting a directory, remote resource, or bulk modification that cannot easily be reverted

- Confirmation:

  - Prompt for confirmation.

  - Provide a **dry run** operation. A **dry run** operation is used to see the results of the operation without actually making any modifications to the data.

- **Severe**:

  - Example: deleting something complex, such as an entire remote application or server

  - Confirmation:

    - Prompt for confirmation along with asking them to either type something non-trivial, such as the name of the resource they are deleting, or use a flag such as `-confirm="name-of-resource"` so it is still scriptable

In general, we want to make it increasingly more difficult for the user to do something more difficult. It is a way of guiding the user away from any accidents.

Any user input should always be validated early on to prevent anything unnecessarily bad from happening. Make the error returned understandable to the user who passed in bad data.

In a conversation, any confidential information must be secured. Make sure that any passwords are protected and provide secure methods for users to submit their credentials. For example, consider only accepting sensitive data via files only. You can offer a `-password-file` flag that allows the user to pass in a file or data via standard input. This method provides a discreet method for passing in secret data.

Be transparent in conversation. Any actions that cross the boundaries of the program should be stated explicitly. This includes reading or writing files that the user did not pass in as arguments unless these files are storing an internal state within a cache. This may also include any actions when talking to a remote server.

Finally, response time is more important than speed. Print something to the user in under 100 milliseconds. If you are making a network request, print out something before the request is made so it doesn't look like the application is hanging or appearing broken. This will make your application appear more robust to its end user.

Let's revisit our audio metadata CLI project. Under *Chapter 8*'s `audiofile` repo, we'll make some changes to create a conversational flow where it might be missing.

Example 1: Prompt for information when a flag is missing

Using the Cobra CLI, if a flag is required, it would automatically return an error if the flag were missing when the command is called. Based on some of the guidelines mentioned in this section, rather than just returning an error, let's prompt for missing data instead. In the audiofile code for *Chapter 8*, in the utils/ask.go file, we create two functions using the survey package github.com/AlecAivazis/survey/v2 as follows:

```
func AskForID() (string, error) {
  id := ""
  prompt := &survey.Input{
    Message: "What is the id of the audiofile?",
  }
  survey.AskOne(prompt, &id)
  if id == "" {
    return "", fmt.Errorf("missing required argument: id")
  }
  return id, nil
}


func AskForFilename() (string, error) {
  file := ""
  prompt := &survey.Input{
    Message: "What is the filename of the audio to upload
      for metadata extraction?",
    Suggest: func(toComplete string) []string {
      files, _ := filepath.Glob(toComplete + "*")
      return files
    },
  }
  survey.AskOne(prompt, &file)
  if file == "" {
    return "", fmt.Errorf("missing required argument:
      file")
  }
  return file, nil
}
```

These two functions can now be called when checking the flags that are passed and whether the values are still empty. For example, in the `cmd/get.go` file, we check for the `id` flag value and if it's still empty, prompt the user for the `id`:

```go
id, _ := cmd.Flags().GetString("id")
if id == "" {
  id, err = utils.AskForID()
  if err != nil {
    return nil, err
  }
}
```

Running this gives the user the following experience:

```
mmontagnino@Marians-MBP audiofile % ./bin/audiofile get
? What is the id of the audiofile?
```

Similarly, in the `cmd/upload.go` file, we check for the filename flag value and if it's still empty, prompt the user for the filename. Because the prompt allows the user to drill down suggested files, we now get the following experience:

```
mmontagnino@Marians-MBP audiofile % ./bin/audiofile upload
? What is the filename of the audio to upload for metadata
extraction? [tab for suggestions]
```

Then, press the Tab key for suggestions to reveal a drill-down menu:

```
mmontagnino@Marians-MBP audiofile % ./bin/audiofile upload
? What is the filename of the audio to upload for metadata
extraction? audio/beatdoctor.mp3 [Use arrows to move, enter to
select, type to continue]
  audio/algorithms.mp3
> audio/beatdoctor.mp3
  audio/nightowl.mp3
```

Providing a prompt helps to guide the user and for them to understand how to run the command works.

Example 2: Confirm deletion

Another way we can help to guide users toward safely using the CLI and protecting them from making any mistakes is to ask the user for confirmation when doing something dangerous. Although it is not necessary to do so during an explicit delete operation, we created a confirmation function that can be used with a configurable message in any type of dangerous situation. The function exists under the `utils/confirm.go` file:

```go
func Confirm(confirmationText string) bool {
  confirmed := false
  prompt := &survey.Confirm{
    Message: confirmationText,
  }
  survey.AskOne(prompt, &confirmed)
  return confirmed
}
```

Example 3: Notify users when making a network request

Before any HTTP request is made, notifying the user helps them to understand what's going on, especially if the request hangs or becomes unresponsive. We've added a message prior to each network request in each command. The `get` command now has the following line prior to the client running the `Do` method:

```go
fmt.Printf("Sending request: %s %s %s...\n",
           http.MethodGet, path, payload)
resp, err := client.Do(req)
if err != nil {
  return nil, err
}
```

Empathy

There are some simple modifications you can make to your command-line application to empathize with your users:

- Be helpful:
  - Provide help text and documentation
  - Suggest commands
  - Rewrite errors in an understandable way
- Invite user feedback and bug submission

In *Chapter 9, The Empathic Side of Development*, we will go through the ways in which you can help guide your users toward success using help text, documentation, widespread support, and providing an effortless way for users to provide feedback and submit bugs.

Example 1: Offering command suggestions

The Cobra CLI offers some empathy when a user mistypes a command. Let's look at the following example where the user mistypes upload as upolad:

```
mmontagnino@Marians-MacBook-Pro audiofile % ./bin/audiofile
upolad
Error: unknown command "upolad" for "audiofile"

Did you mean this?
        upload


Run 'audiofile --help' for usage.
```

Example 2 – Offer an effortless way to submit bugs

In *Chapter 9, The Empathic Side of Development,* we define a bug command that will launch the default browser and navigate to the GitHub repository's new issue page to file a bug report:

```
mmontagnino@Marians-MacBook-Pro audiofile % ./bin/audiofile bug
--help
Bug opens the default browser to start a bug report which will
include useful system information.

Usage:
  audiofile bug [flags]

Examples:
audiofile bug
```

Example 3: Print usage command is used incorrectly

Suppose a user does not input a value to search for when running the search command. The CLI application will prompt for a value to search for. If a value is not passed in by the user, the CLI will output the proper usage of the command:

```
mmontagnino@Marians-MacBook-Pro audiofile % ./bin/audiofile
search
? [Q] What value are you searching for?
```

```
Error: missing required argument (value)
Usage:
  audiofile search [flags]


Flags:
  -h, --help            help for search
      --json            return json format
      --plain           return plain format
      --value string    string to search for in metadata
```

Personalization

In general, make the default the right thing for most users, but also allow users to personalize their experience with your CLI. The configuration gives the users a chance to personalize their experience with your CLI and make it more their own.

Example 1: Technical configuration with Viper

Using `audiofile` as an example, let's create a simple configuration setup with Viper to offer the user the ability to change any defaults to their liking. The configurations that we've created are for the API and CLI applications. For the API, we've defined the `configs/api.json` file, which contains the following:

```
{
  "api": {
    "port": 8000
  }
}
```

The API will always execute locally to where it's being executed. Then, for the CLI, we've defined a similar simple file, `configs/cli.json`, containing the following:

```
{
  "cli": {
    "hostname": "localhost",
    "port": 8000
  }
}
```

If the API is running on an external host with a different port, then these values can be modified within the configuration. For the CLI to point to the new hostname, we'll need to update any references within the CLI commands to use the value in the configuration. For example, in the cmd/get.go file, the path is defined as:

```
path := fmt.Sprintf("http://%s:%d/request?%s",
    viper.Get("cli.hostname"), viper.GetInt("cli.port"),
    params)
```

To initialize these values and provide defaults if any required values are missing from the configuration, we run a Configure function defined in cmd/root.go:

```
func Configure() {
  viper.AddConfigPath("./configs")
  viper.SetConfigName("cli")
  viper.SetConfigType("json")
  viper.ReadInConfig()
  viper.SetDefault("cli.hostname", "localhost")
  viper.SetDefault("cli.port", 8000)
}
```

A similar code exists within the cmd/api.go file to gather some of the same information. Now that this is set up, if there are any changes the user wants to make to the hostname, log level, or port, there is only one configuration file to modify.

Example 2: Environment variable configuration

Suppose there is an environment variable specific to the application that allows users to define the foreground and background color to use. This environment variable could be named AUDIOFILE_ COLOR_MODE. Using the Viper configuration again, values for the foreground and background texts may be used to overwrite default settings. While this is not implemented within our CLI, the Viper configuration may look like the following:

```
{
  "cli": {
    "colormode": {
      "foreground": "white",
      "background": "black",
    }
  }
}
```

Example 3: Storage location

Sometimes users want the location of certain output, logging, for example, to be stored in a particular area. Providing details within Viper can allow defaults to be overwritten. Again, this is not currently implemented within our CLI, but if we were to provide this option within our configuration, it may look like this:

```
{
  "api": {
    "local_storage": "/Users/mmontagnino/audiofile"
  }
}
```

Any other new configuration values can be added with a similar approach. Providing the ability to configure your application is the start for personalization. Think of the many ways you can configure your CLI: color settings, disabling prompts or ASCII art, default formatting, and more.

Pagination

Use a pager when you are outputting a lot of text, but be careful because sometimes the implementation can be error-prone.

Pagination for Unix or Linux

On a Unix or Linux machine, you may use the `less` command for pagination. Calling the `less` command with a sensible set of options, such as `less -FIRX`, pagination does not occur if the contents fit on a single screen, case is ignored when searching, color and formatting are enabled, and the content is kept on the screen when `less` quits. We will use this as an example within the next section when outputting table data, and in preparation, within the `utils` package, we add the following files: `pager_darwin.go` and `pager_linux.go`, with a `Pager` function. In our case, though, we use the `-r` flag only because we want to continue displaying colors in the table:

```go
func Pager(data string) error {
  lessCmd := exec.Command("less", "-r")
  lessCmd.Stdin = strings.NewReader(data)
  lessCmd.Stdout = os.Stdout
  lessCmd.Stderr = os.Stderr
  err := lessCmd.Run()
  if err != nil {
    return err
  }
```

```
        return nil
    }
```

Pagination for Windows

On a Windows machine, we use the `more` command instead. Within the `utils` package, we add the `pager_windows.go` file following with a `Pager` function:

```
func Pager(data string) error {
    moreCmd := exec.Command("cmd", "/C", "more")
    moreCmd.Stdin = strings.NewReader(data)
    moreCmd.Stdout = os.Stdout
    moreCmd.Stderr = os.Stderr
    err := moreCmd.Run()
    if err != nil {
        return err
    }
    return nil
}
```

Now you know how to handle the pagination of output on the three major operating systems. This will also help users when you are outputting a large amount of data to scroll through the output easily.

Visual language

Depending on the data, it might be easier for the users to see it in plain text, table format, or in JSON format. Remember to provide the user with options to return data in the format they prefer with the `-plain` or `-json` flag.

> **Note**
> Sometimes, for all of the data to appear within a user's window, some lines may be wrapped within a cell. This will break scripts.

There are many visual cues that can be displayed to the user to increase information density. For example, if something is going to take a long time, use a progress bar and provide an estimate of the time remaining. If there is a success or failure, utilize color codes to provide an additional level of information for the user to consume.

We now know how to determine whether we are outputting to a human via a terminal or to another application, so knowing the difference allows us to output data appropriately. Let's continue to the next section to discuss fun examples to provide data with ASCII visualizations to improve information density.

Increasing information density with ASCII art

As the title of this section states, you can increase information density using ASCII art. For example, running the `ls` command shows file permissions in a way a user can easily scan with their eyes and understand with pattern recognition. Also, using a highlighter pen when studying in a textbook to literally highlight a sentence or group of words makes certain phrases jump out as more important. In this section, we'll talk about some common uses for ASCII art to increase the understanding of the importance of shared information.

Displaying information with tables

Probably the clearest way that data can be displayed to users is in a table format. Just like the `ls` format, patterns can jump out more easily in a table format. Sometimes records can contain data that is longer than the width of the screen and lines become wrapped. This can break scripts that might be relying on one record per line.

Let's take our audiofile as an example and instead of returning the JSON output, use the package to return the data cleanly in a table. We can keep the ability to return JSON output for when the user decides to require it using the `-json` flag.

The simplest way of outputting data as a table with the `pterm` package is using the default table. Next to the models, there currently exists a `JSON()` method that will take the struct and then output it in JSON format. Similarly, we add a `Table()` method on the pointer to the struct. In the `models/audio.go` file, we add the following bit of code for the header table:

```
var header = []string{
  "ID",
  "Path",
  "Status",
  "Title",
  "Album",
  "Album Artist",
  "Composer",
  "Genre",
  "Artist",
  "Lyrics",
  "Year",
```

```
      "Comment",
    }
```

This defines the header for the audio table. We then add some code to transform an `audio` struct into a row:

```
func row(audio Audio) []string {
  return []string{
    audio.Id,
    audio.Path,
    audio.Status,
    audio.Metadata.Tags.Title,
    audio.Metadata.Tags.Album,
    audio.Metadata.Tags.AlbumArtist,
    audio.Metadata.Tags.Composer,
    audio.Metadata.Tags.Genre,
    audio.Metadata.Tags.Artist,
    audio.Metadata.Tags.Lyrics,
    strconv.Itoa(audio.Metadata.Tags.Year),
    strings.Replace(audio.Metadata.Tags.Comment, "\r\n",
        "", -1),
  }
}
```

Now we use the `pterm` package to create the table from the header row and function to convert an audio item into a row, each of type `[]string`. The `Table()` method for `Audio` and `AudioList` structs are defined below:

```
func (list *AudioList) Table() (string, error) {
  data := pterm.TableData{header}
  for _, audio := range *list {
    data = append(
      data,
      row(audio),
    )
  }
  return pterm.DefaultTable.WithHasHeader()
      .WithData(data).Srender()
```

```go
}

func (audio *Audio) Table() (string, error) {
  data := pterm.TableData{header, row(*audio)}
  return pterm.DefaultTable.WithHasHeader().WithData(data).
    Srender()
}
```

All the data in this example is output one record per line. If you decide on a different implementation and this is not the case for your code, make sure you add the -plain flag as an optional flag where once it is called, it will print one record per line. Doing this will ensure that scripts do not break on the output of the command. Regardless, depending on the size of the data and terminal, you may notice that the data wraps around and might be hard to read. If you are running Unix, run the tput rmam command to remove line wrapping from terminal.app and then tput smam to add line wrapping back in. On Windows, there will be a setting under your console properties. Either way, this should make viewing the table data a bit easier!

If a lot of data is returned within the table, then it's important to add paging for increased usability. As mentioned in the last section, we've added a Pager function to the utils package. Let's modify the code so that it checks whether the data is being output to a terminal, and if so, page the data using the Pager function. In the utils/print.go file, within the Print function, we paginate the JSON formatted data, for example, as follows:

```go
if jsonFormat {
    if IsaTTY() {
        err = Pager(string(b))
        if err != nil {
            return b, fmt.Errorf("\n paging: %v\n ", err)
        }
    } else {
        return b, fmt.Errorf("not a tty")
    }
}
```

If the output is returned to a terminal, then we paginate, otherwise we return the bytes with an error that informs the calling function it is not a terminal. For example, the cmd/list.go file calls the preceding Print function:

```
formatedBytes, err := utils.Print(b, jsonFormat)
if err != nil {
    fmt.Fprintf(cmd.OutOrStdout(), string(formatedBytes))
}
```

When it receives the error, then it just prints the string value to standard output.

Clarifying with emojis

A picture is worth a thousand words. So much information can be shared just by adding an emoji. For example, think of the simple green checkbox, ✅, that is so often used on Slack or in GitHub to signal approval. Then, there is the opposite case with a red x, ✖, to symbolize that something went wrong.

Emojis are letters that exist within the UTF-8 (Unicode) character set, which covers almost all the characters and symbols of the world. There are websites that will share this Unicode emoji mapping. Visit https://unicode.org/emoji/charts/full-emoji-list.html to view the full character list.

Example 1 – Green checkmark for successful operations

In our audiofile, we add the emoji to the output to the upload command. At the top of the file, we add the emoji constant with a UTF-8 character code:

```
const (
    checkMark = "\U00002705"
)
```

Then, we use it in the following output:

```
fmt.Println(checkMark, " Successfully uploaded!")
fmt.Println(checkMark, " Audiofile ID: ", string(body))
```

Running the upload command after a new recompile and run shows the emoji next to the output, indicating a successful upload. The green checkmark assures the user that everything ran as expected and that there were no errors:

```
✅ Successfully uploaded!
✅ Audiofile ID: b91a5155-76e9-4a70-90ea-d659c66d39e2
```

Example 2 – Magnifying glass for search operations

We've also added a magnifying glass, 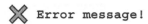, in a similar way when the user runs the search command without the `--value` flag. The new prompt looks like this:

? 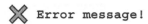 **What value are you searching for?**

Example 3 – Red for error messages

If there is an invalid operation or an error message, you could also add a red x to symbolize when something goes wrong:

Error message!

Emojis not only add a fun element to your CLI but also a very valuable one. The little emoji is another way to increase information density and get important points across to the user.

Using color with intention

Adding color highlights important information for the end user. Don't overdo it, though; if you end up with multiple different colors frequently used throughout, it's hard for anything to jump out as important. So use it sparingly, but also intentionally.

An obvious color choice for errors is red, and success is green. Some packages make adding color to your CLI easy. One such package we will use in our examples is `https://github.com/fatih/color`.

Within the audiofile, we look at a few examples where we could integrate colors. For example, the ID for the table that we just listed out. We import the library and then use it to change the color of the ID field:

```
var IdColor = color.New(color.FgGreen).SprintFunc()
func row(audio Audio) []string {
  return []string{
    IdColor(audio.Id),
    ...
  }
}
```

In the `utils/ask.go` file, we define an `error` function that can be used within the three ask prompts:

```
var (
  missingRequiredArumentError =
    func(missingArg string) error {
    return fmt.Errorf(errorColor(fmt.Sprintf("missing
      required argument (%s)", missingArg)))
  }
)
```

The `fmt.Errorf` function receives the `errorColor` function, which is defined within a new `utils/errors.go` file:

```
package utils

import "github.com/fatih/color"

var errorColor = color.New(color.BgRed,
  color.FgWhite).SprintFunc()
```

Together, we recompile code and try to run it again, purposely omitting required flags from commands. We see that the command errors out and prints the error with a red background and white foreground, defined by the `color.BgRed` and `color.FgWhite` values. There are many ways to add color. In the `color` package we're using, the prefix `Fg` stands for foreground and the prefix `Bg` stands for background.

Use colors intentionally, and you will visually transfer the most important information easily to the end user.

Spinners and progress bars

Spinners and progress bars signify that the command is still processing; the only difference is that progress bars visually display progress. Since it is common to build concurrency into applications, you can also show multiple progress bars running simultaneously. Think about how the Docker CLI often shows multiple files being downloaded simultaneously. This helps the user understand that there's something happening, progress is made, and nothing is stalling.

Example 1 – Spinner while playing music

There are different ways that you can add spinners to your Golang project. In the audiofile project, we'll show a quick way to add a spinner using the `github.com/pterm/pterm` package. In the audiofile project, for each play command distinct for each operating system, we add some code to start and stop the spinner. Let's look at `play_darwin.go`, for example:

```go
func play(audiofilePath string) error {
    cmd := exec.Command("afplay", audiofilePath)
    if err := cmd.Start(); err != nil {
        return err
    }
    spinnerInfo := &pterm.SpinnerPrinter{}
    if utils.IsaTTY() {
        spinnerInfo, _ = pterm.DefaultSpinner.Start("Enjoy the
            music...")
    }
    err := cmd.Wait()
    if err != nil {
        return err
    }
    if utils.IsaTTY() {
        spinnerInfo.Stop()
    }
    return nil
}
```

Running the `play` command for any audio file shows the following output:

■ **Enjoy the music... (3m54s)**

It's hard to capture the spinner in the previous line, but the black box spins around in a circle while the music plays.

Example 2 – Progress bar when uploading a file

Next, within the `upload` command, we can show code to display the progress of uploading a file. Since the API only uses local flat file storage, the upload goes so quickly it's hard to see the change in the progress bar, but you can add some `time.Sleep` calls in between each increment to see the progress appear more gradually. Within the `cmd/upload.go` file, we've added several statements to create the progress bar and then increment the progress along with title updates:

```
p, _ := pterm.DefaultProgressbar.WithTotal(4).
WithTitle("Initiating upload...").Start()
```

This first line initiates the progress bar, and then to update the progress bar, the following lines are used:

```
pterm.Success.Println("Created multipart writer")
p.Increment()
p.UpdateTitle("Sending request...")
```

Notice that when we first define the progress bar, we call the `WithTotal` method, which takes the total number of steps. This means that for each step where `p.Increment()` is called, the progress bar progresses by 25 percent or 100 divided by the total number of steps. When running a spinner, it's great to add the visualizer to let the user know that the application is currently running a command that might take a while:

```
Process response... [4/4]  ███████████             65% | 5s
```

The progress bar gives the user a quick visual of how quickly the command is progressing. It's a great visual indicator for any command that will take a long time and can be clearly split into multiple steps for progression. Again, spinners and progress bars should not be displayed unless the output is being displayed to the terminal or TTY. Make sure you add a check for TTY before outputting the progress bar or spinner.

Disabling colors

There are different reasons why color may be disabled for a CLI. A few of these things include:

- The standard out or standard error pipe is not connected to a TTY or interactive terminal. There is one exception to this. If the CLI is running within a CI environment, such as Jenkins, then color is usually supported, and it is recommended to keep color on.

- The `NO_COLOR` or `MYAPP_NO_COLOR` environment variable is set to true. This can be defined and set to disable color for all programs that check it or specifically for your program.

- The TERM environment variable is set to dumb.

- The user passes in the -no-color flag.

Some percentage of your users may be colorblind. Allowing your users to swap out one color for another is a nice way to consider this specific part of your user base. This could be done within the configuration file or application. Allowing them to specify a color and then overwrite it with a preferred color will again allow the user to customize the CLI. This customization will provide users with an improved experience.

Including ASCII art within your application increases information density—a visual indicator that easily helps users to understand some important information. It adds clarity and conciseness. Now let's discuss a way to make your CLI more intuitive through consistency.

Being consistent across CLIs

Learning about command-line syntax, flags, and environment variables requires an upfront cost that pays off in the long run with efficiency if programs are consistent across the board. For example, terminal conventions are ingrained into our fingertips. Reusing these conventions by following preexisting patterns helps to make a CLI more intuitive and guessable. This is what makes users efficient.

There are times when preexisting patterns break usability. As mentioned earlier, a lot of Unix commands don't return any output by default, which can cause confusion for people who are new to using the terminal or CLI. In this case, it's fine to break the pattern for the benefit of increased usability.

There are specific topics to consider when maintaining consistency with the larger community of CLIs, but also within the application itself:

- Naming

- Positional versus flag arguments

- Flag naming

- Usage

Naming

Use consistent command, subcommand, and flag names to help users intuit your command-line application. Some modern command-line applications, such as the AWS command-line application, will use Unix commands to stay consistent. For example, look at this AWS command:

```
aws s3 ls s3://mybucket --summarize
```

The previous command uses the `ls` command to list `S3` objects in the `S3` bucket. It's important to use common, and non-ambiguous, command names outside of reusing shell commands in your CLI. Take the following as examples that can be logically grouped by type:

Management	Development	Server	Other
create	build	start	help
get	deploy	stop	version
update	import		
delete	init		
list	push		
search	test		

Table 8.1 – Example grouping commands by type

These are common names across CLIs. You can also consider integrating some common Unix commands:

- `cp` (copy)
- `ls` (list)
- `mv` (move)

These common command names remove confusion from a long list of ambiguous or unique names. One common confusion is the difference between the update and upgrade commands. It's best to use one or the other as keeping both will only confuse your users. Also, for the command names that are used often, follow the standard shorthand for these popular commands as well. For example:

- `-v, --version`
- `-h, --help`
- `-a, --all`
- `-p, --port`

Rather than listing all examples, just consider some of the most common command-line applications you use. Think about which command names make sense for consistency across the board. This will benefit not only your application but the community of command-line applications as a whole as further standards are solidified.

Positional versus flag arguments

It's important to stay consistent with arguments and their position. For example, in the AWS CLI, the s3 argument is consistently next to its arguments:

```
aws s3 ls s3://<target-bucket>
aws s3 cp <local-file> <s3-target-location>/<local-file>
```

The consistent position of specific arguments will build a clear pattern that users will follow intuitively.

If flags, that we had mentioned before, are available with one command, they can be available for another command where they make sense. Rather than changing the flag name for each command, stay consistent between commands. Do the same with subcommands. Let's look at some examples from the GitHub CLI:

```
gh codespace list --json
gh issue list -json
```

The GitHub CLI keeps the list subcommand consistent across different commands and reuses the -json flag, which has the same behavior across the application.

> **Note**
> Required arguments are usually better as positional rather than flags.

Flag naming

Not only is it important to stay consistent on the position of arguments and the flag names across different commands, but it's also important to be consistent within the naming. For example, there are flags that can be defined in camel case, -camelCase, snake case, --SnakeCase, or with dashes, --flag-with-dashes. Staying consistent with the way you are naming your flags in your application is also important!

Usage

In previous chapters, we discussed the grammar of a command and how applications can be defined with a consistent structure: **noun-verb** or **verb-noun**. Staying consistent with the structure also lends to a more intuitive design.

When building your command-line application, if you think about how to stay consistent across other programs and internal to your application, you will create a more intuitive and easier to learn command-line application where your users feel naturally supported.

Summary

In this chapter, you learned some specific points to consider when building for a machine versus a human. Machines like simple text and have certain expectations of the data that is returned from other applications. Machine output can sometimes break usability. Designing for humans first, we talked about how we can easily switch to machine-friendly output when needed with the use of some popular flags: `--json`, `--plain`, and `--silence`.

Much goes into a usable design, and we went over some of the ways you can increase the usability of your CLI—from using color with intention, outputting data in tables, paging through long text, and being consistent. All of the aforementioned elements will help the user feel more comfortable and guided when using your CLI, which is one of the main goals we want to achieve. We can summarize with a quick table what a good CLI design looks like versus a bad CLI design:

CLI Design	
Good	**Bad**
Checks if outputting to a human versus another machine process	Outputs the same to human versus another machine process
Allows user to change output format	No options to change output format
Allows user to quiet unnecessary output	No silence options
Uses conversational text for humans	Sends conversational text, emojis, and color output to another machine process
Outputs plain text for machine processes	Sends no output at all
Validates input early	No input validation
Transparent with users	Local file modification and network requests without notifying the user
Prioritizes response time over speed	Long response time
Prompts for missing input	No prompts for required input
Confirmation before irreversible deletions	No confirmation before major modifications
Notify user of network requests	Unclear what is happening
Empathic text	Unempathetic text
Use of intentional visual language	Many colors used and meaning is lost
Allows personalization	No personalization
Pagination for long text output	No pagination on long text output
Progress bars/spinners to show processing	No progress bar to indicate processing
Importing information at the bottom	Importing information at the top of output or missing

Figure 8.5 – Good versus bad CLI design

In the next chapter, *Chapter 9, Empathic Side of Development*, we will continue discussing how to develop for humans by incorporating more empathy.

Questions

1. What common flags in scripts can be used with a command-line application to keep the output stable?

2. What flag should you check to see if the end user does not want color set within the terminal? And what common flag can be used to disable color from the output?

3. Think about how there could be two commands with similar names and how this adds ambiguity. What ambiguous commands have you come across in your experience of CLIs?

Further reading

*   *The Anti-Mac Interface*: `https://www.nngroup.com/articles/anti-mac-interface/`

*   *The Humane Interface: New Directions for Designing Interactive Systems* by Jef Raskin

Answers

1. `--json` and `--plain` flags keep data consistent and reduce the risk of breaking scripts.

2. Either the `TERM=dumb`, `NO_COLOR`, or `MYAPP_NO_COLOR` environment variables. The most common flag for disabling color is the `–no-color` flag.

3. Update versus upgrade are commonly confused, as well as name and host.

9

The Empathic
Side of Development

Empathy has been a hot topic lately, and its relation to software is no exception. This chapter will discuss how to use empathy to develop a better CLI. Empathy-driven CLI development is done with consideration of the output and errors that are written and the clarity and reassurance it may give the user. Written documentation that takes an empathetic approach also provides users with an effortless way to get started, while help and support are readily available for users when they need it.

This chapter will give examples of how to rewrite errors in a way that users may easily understand, not just by being clearer that an error occurred but also how and where (with debug and traceback information), which can be provided with a `--verbose` flag and detailed logging. It is very important to provide logs for users, and this implementation will be described when discussing debug and traceback information. Users can also feel more reassured with the help of man pages, usage examples of each command, empathically written documentation, and a quick and easy way to submit bugs that are encountered within the application.

Taking an empathetic approach into many different areas of your application, as well as in your life, is a form of not only self-care but care for others as well. Hopefully, these tips will help to create a CLI that meets the user at their perspective and provides them with a feeling of reassurance. Specifically, this chapter will cover the following topics:

- Rewriting errors to be human-readable
- Providing debug and traceback information
- Effortless bug submission
- Help, documentation, and support

Technical requirements

These are the requirements for this chapter:

- A Unix operating system to understand and run the examples shared in the chapter
- You can also find the code examples on GitHub at `https://github.com/PacktPublishing/Building-Modern-CLI-Applications-in-Go/tree/main/Chapter09`

Rewriting errors to be human-readable

Errors can be a big point of frustration for users as they can set users off their original plans. Users will be grateful, though, if you can make the process as painless as possible. In this section, we will discuss some ways to ease users when an error occurs and provide some guidelines for creating better error messages and avoiding some common mistakes. Creating clear and helpful error messages is often overlooked, yet they are very impactful toward an optimal UX.

Think of some of your subjective experiences while working with CLIs and some of the errors you have encountered. This is an opportunity to think about how experiences can be improved for yourself when working with your own CLI, but also for others.

Guidelines for writing error messages

Here are some useful guidelines when writing error messages:

- **Be specific**: Customize messages toward the actual task that has occurred. This error message is critical if the task required inputting credentials or a final command to complete a workflow. The best experience would include specifying the exact problem and then providing a way toward correcting the issue. Specific guidance helps the users stay engaged and willing to make corrections.

- **Remind the user that there's a human on the other end**: A generic error message can sound very technical and intimidating to most users. By rewriting the error message, you can make them more useful and less intimidating. Empathize with your users and make sure that you don't place blame on the user, which can be particularly discouraging. It's important to encourage the user by being understanding, friendly, and speaking the same language, both literally and figuratively! How do the words you use sound in conversation?

- **Keep it light-hearted**: Keeping a light-hearted tone can help ease any tension when an error occurs, but be careful! In certain situations, it might make the situation a bit worse—especially if it's a critical task. Users do not want to feel mocked. Regardless, with humor or not, the error message should still be informational, clear, and polite.

- **Make it easy**: This will require you to do a bit more of the heavy lifting, but it will certainly be worth it in the end. Provide clear next steps, or commands to run, to resolve the issue and to help the user get back on track to what they had originally intended on doing. With helpful suggestions, the user will at least see the path through the trees and know what to do next.

- **Consider the best placement**: When outputting error messages, it's best to place them in an area where users will look first. In the case of the CLI, it's most likely at the end of the output.

- **Consolidate errors**: If there are multiple error messages, especially similar ones, group them together. It will look much better than repeating the same error message over and again.

- **Optimize your error message with icons and text**: Usually, important information is placed at the end of the output, but if there's any red text on the screen, that is often where the user's eyes will be drawn to. Given the power of color, use it sparingly and with intention.

- **Consider capitalization and punctuation**: Don't write in all caps or with multiple exclamation points. Consider consistency as well—do your errors start with capitalization? If they are output to a log, errors may start all in lowercase letters.

Decorating errors

Wrapping errors with additional information and context is a very important step. What is the specific task that failed and why? This helps the user understand what happened. Providing actions to take toward resolution will also help the user feel more supported and willing to move forward.

First, there are a few ways to decorate your errors with additional information. You can use the fmt. Errorf function:

```
func Errorf(format string, a ...interface{}) error
```

With this function, you can print out the error as a string with any additional context. Here's an example within the errors/errors.go file in the Chapter-9 repo:

```
birthYear := -1981
err := fmt.Errorf("%d is negative\nYear can't be negative",
birthYear)
if birthYear < 0 {
    fmt.Println(err)
} else {
    fmt.Printf("Birth year: %d\n", birthYear)
}
```

The next way to decorate your errors is by using the `errors.Wrap` method. This method is fully defined as follows:

```
func Wrap(err error, message string) error
```

It returns an error annotating `err` with a message and a stack trace at the point the method is called. If `err` is `nil`, then the `Wrap` function also returns `nil`.

In the `wrapping()` function, we demonstrate this:

```
func wrapping() error {
    err := errors.New("error")
    err1 := operation1()
    if err1 != nil {
        err1 = errors.Wrap(err, "operation1")
    }
    err2 := operation2()
    if err != nil {
        err2 = errors.Wrap(err1, "operation2")
    }
    err3 := operation3()
    if err != nil {
        err3 = errors.Wrap(err2, "operation3")
    }
    return err3
}
```

Notice that the previous error gets wrapped into the next error and so on until the final error is returned. The output of the error returned from the `wrapping()` function is shown here. I've removed the longer path for clarity:

```
error
.../errors.wrapping
        .../errors/errors.go:73
.../errors.Examples
        .../errors/errors.go:39
main.main
        .../main.go:6
```

```
runtime.main
        /usr/local/go/src/runtime/proc.go:250
runtime.goexit
        /usr/local/go/src/runtime/asm_amd64.s:1594
operation1
.../errors.wrapping
        .../errors/errors.go:76
.../errors.Examples
        .../errors/errors.go:39
main.main
        .../main.go:6
runtime.main
        /usr/local/go/src/runtime/proc.go:250
runtime.goexit
        /usr/local/go/src/runtime/asm_amd64.s:1594
operation2
.../errors.wrapping
        .../errors/errors.go:80
.../errors.Examples
        .../errors/errors.go:39
main.main
        .../main.go:6
runtime.main
        /usr/local/go/src/runtime/proc.go:250
runtime.goexit
        /usr/local/go/src/runtime/asm_amd64.s:1594
operation3
.../errors.wrapping
        .../errors/errors.go:84
.../errors.Examples
        .../errors/errors.go:39
main.main
        .../main.go:6
runtime.main
        /usr/local/go/src/runtime/proc.go:250
```

```
runtime.goexit
        /usr/local/go/src/runtime/asm_amd64.s:1594
```

Notice that the errors from `operation1`, `operation2`, and `operation3` are wrapped under the original `error` instance.

Because wrapping annotates the error with the stack trace and message, the line that calls the `wrapping()` function prints the error message followed by the stack trace at the call of the `New()` or `Wrap()` method.

Customizing errors

Creating custom errors allows you to store whatever information you think is valuable to your users with the error so that when it's time to print out, all the information is available within a single struct. First, you need to think about the error structure:

```
type error interface {
    Error() string
}
```

Simply create any type that implements the `Error()  string` method. Think about the data you'd want stored on the custom error structure that might be useful for your users, or even for yourself as the developer, for debugging purposes. This could include the method name where the error occurred, the severity of the error, or the kind of error. In the `Chapter-9` repo, in the `errors.go` file, I provide some examples. To keep things simple, only one additional field, `Task`, is added to the `customError` structure:

```
type customError struct {
    Task string
    Err error
}
```

The `Error()` method that satisfies the previous interface is defined here. For fun, we use the `github.com/fatih/color` color page used in the previous chapter and an emoji (a red cross mark) alongside the error message:

```
func (e *customError) Error() string {
    var errorColor = color.New(color.BgRed,
        color.FgWhite).SprintFunc()
    return fmt.Sprintf("%s: %s %s", errorColor(e.Task),
```

```
        crossMark, e.Err)
    }
```

Now, we can demonstrate how this custom error can be used within the `eligibleToVote` function:

```
func eligibleToVote(age int) error {
    fmt.Printf("%s Attempting to vote at %d years
        old...\n", votingBallot, age)
    minimumAge := 18
    err := &customError{
        Task: " eligibleToVote",
    }
    if age < minimumAge && age > 0 {
        years := minimumAge - age
        err.Err = fmt.Errorf("too young to vote, at %d,
            wait %d more years", age, years)
        return err
    }
    if age < 0 {
        err.Err = fmt.Errorf("age cannot be negative: %d",
            age)
        return err
    }
    fmt.Println("Voted.", checkMark)
    return nil
}
```

Notice there are multiple errors, and the error is initially defined at the top of the function, setting only the `Task` field. For each error that occurs, the `Err` field is then set and returned. Within the `Examples` method, we call the function with the following lines:

```
birthYear = 2010
currentYear := 2022
age := currentYear - birthYear
err = eligibleToVote(age)
if err != nil {
    fmt.Println("error occurred: ", err)
}
```

The following error is output when the preceding code runs:

```
Attempting to vote at 12 years old...
error occurred:  eligibleToVote: X too young to vote, at 12, wait 6 more years
```

Figure 9.1 – Screenshot of voting error

There are plenty of other ways to create custom errors, but here are a few things to consider adding to your custom errors:

- The severity of the error for logging purposes

- Any data that may be valuable for metrics

- The kind of error so that you may easily filter out any unexpected errors when they occur

Writing better error messages

Now that we know how to add more detail to error messages, let's revisit the audiofile CLI and rewrite our error messages to be more human-friendly using the guidelines mentioned earlier in this section. In the repo, for this particular branch, I've decorated the errors with extra information so that the user or developer can better understand where the error occurred and why.

Since the audiofile CLI interacts with the audiofile API, there are HTTP responses that can be handled and rewritten. A CheckResponse function exists in the utils/http.go file and does this:

```
func CheckResponse(resp *http.Response) error {
    if resp != nil {
        if resp.StatusCode != http.StatusOK {
            switch resp.StatusCode {
            case http.StatusInternalServerError:
                return fmt.Errorf(errorColor("retry the command
                    later"))
            case http.StatusNotFound:
                return fmt.Errorf(errorColor("the id cannot be
                    found"))
            default:
                return fmt.Errorf(errorColor(fmt.
                    Sprintf("unexpected response: %v", resp.
                    Status)))
            }
        }
        return nil
```

```
    } else {
        return fmt.Errorf(errorColor("response body is nil"))
    }
}
```

Consider expanding on this within your own CLI, which might also interact with a REST API. You may check as many responses as you like and rewrite them as errors to be returned by the command.

In previous versions of the audiofile CLI, if an id parameter was passed into the get or delete command, nothing would be returned if the ID was not found. However by passing back the http. StatusNotFound response and adding additional error decorations, the command that would previously error silently and return no data can now return some useful information:

```
mmontagnino@Marians-MacBook-Pro audiofile % ./bin/audiofile get
--id 1234
Sending request: GET http://localhost:8000/request?id=1234 ...
Error:
  checking response: the id cannot be found
Usage:
  audiofile get [flags]

Flags:
  -h, --help          help for get
      --id string     audiofile id
      --json          return json format
```

We can even level up by additionally suggesting how to find an ID. Potentially, ask the user to run the list command to confirm the ID. Another thing that can be done, similarly to how we handled the status codes from an HTTP API request, is to check the errors coming back from a local command being called. Whether the command is not found or the command is missing executable permissions, you can similarly use a switch to handle potential errors that can occur when a command is started or run. These potential errors can be rewritten similarly using more user-friendly language.

Providing debug and traceback information

Debug and traceback information is mostly useful for you or other developers, but it can also help your end users share valuable information with you to help debug a potential issue found in your code. There are several diverse ways to provide this information. Debug and traceback information is primarily output to a log file, and often, the addition of a verbose flag will print this output, which is usually hidden.

Logging data

Since debug data is usually found in log files, let us discuss how to include logging in your command-line application and determine the levels associated with logging—`info`, `error`, and `debug` levels of severity. In this example, let us use a simple log package to demonstrate this example. There are several different popular structured log packages, including the following:

- Zap (`https://github.com/uber-go/zap`)—Fast structured logger developed by Uber
- ZeroLog (`https://github.com/rs/zerolog`)—Fast and simple logger dedicated to JSON format
- Logrus (`https://github.com/sirupsen/logrus`)—Structured logger for Go with the option for JSON-formatted output (currently in maintenance mode)

Although `logrus` is an extremely popular logger, it has not been updated in a while, so let us choose to use `zap` instead. In general, it's a promising idea to choose an open source project that is actively maintained.

Initiating a logger

Back to the `audiofile` project, let us add logging for debugging purposes. The very first thing we run under our `audiofile` repo is this:

```
go get -u go.uber.org/zap
```

It will get the updated Zap logger dependencies. After that, we can start referencing the import within the project's Go files. Under the `utils` directory, we add a `utils/logger.go` file to define some code to initiate the Zap logger, which is called within the `main` function:

```
package utils

import (
    "go.uber.org/zap"
)

var Logger *zap.Logger
var Verbose *zap.Logger

func InitCLILogger() {
    var err error
    var cfg zap.Config
    config := viper.GetStringMap("cli.logging")
```

```
    configBytes, _ := json.Marshal(config)
    if err := json.Unmarshal(configBytes, &cfg); err != nil {
        panic(err)
    }
    cfg.EncoderConfig = encoderConfig()
    err = createFilesIfNotExists(cfg.OutputPaths)
    if err != nil {
        panic(err)
    }
    cfg.Encoding = "json"
    cfg.Level = zap.NewAtomicLevel()
    Logger, err = cfg.Build()
    if err != nil {
        panic(err)
    }
    cfg.OutputPaths = append(cfg.OutputPaths, "stdout")
    Verbose, err = cfg.Build()
    if err != nil {
        panic(err)
    }
    defer Logger.Sync()
}
```

It isn't necessary, but we define two loggers here. One is a logger, `Logger`, which outputs to an output path defined within the config file, and the other is the verbose logger, `Verbose`, which outputs to standard output and the previously defined output path. Both use the `*zap.Logger` type, which is used when type safety and performance are critical. Zap also provides a sugared logger, which is used when performance is nice to have but not critical. `SugarLogger` also allows for structured logging, but in addition, supports `printf`-style APIs.

Within the `Chapter-9` branch version of this repo, we replace some of the general `fmt.Println` or `fmt.Printf` output with the logs that can be shown in `verbose` mode. Also, we differentiate when printing out information with the `Info` level versus the `Error` level.

The following code uses Viper to read from the configuration file, which has been modified to hold a few extra configurations for the logger:

```
{
    "cli": {
```

```
        "hostname": "localhost",
        "port": 8000,
        "logging": {
            "level": "debug",
            "encoding": "json",
            "outputPaths": [
                "/tmp/log/audiofile.json"
            ]
        }
    }
}
```

In the preceding configuration, we set the `level` and `encoding` fields. We choose the `debug` level so that debug and error statements are output to the log file. For the `encoding` value, we chose `json` because it provides a standard structure that can make it easier to understand the error message as each field is labeled. The encoder config is also defined within the same `utils/logger.go` file:

```go
func encoderConfig() zapcore.EncoderConfig {
    return zapcore.EncoderConfig{
        MessageKey: "message",
        LevelKey: "level",
        TimeKey: "time",
        NameKey: "name",
        CallerKey: "file",
        StacktraceKey: "stacktrace",
        EncodeName: zapcore.FullNameEncoder,
        EncodeTime: timeEncoder,
        EncodeLevel: zapcore.LowercaseLevelEncoder,
        EncodeDuration: zapcore.SecondsDurationEncoder,
        EncodeCaller: zapcore.ShortCallerEncoder,
    }
}
```

Since the `InitCLILogger()` function is called within the `main` function, the two `Logger` and `Verbose` loggers will be available within any of the commands for use.

Implementing a logger

Let us look at how we can start using this logger in an effective way. First, we know that we are going to log all the data and output to the user when in verbose mode. We define the `verbose` flag as a persistent flag in the `cmd/root.go` file. This means that the `verbose` flag will be available not only at the root level but also for every subcommand added to it. In that file's `init()` function, we add this line:

```
rootCmd.PersistentFlags().BoolP("verbose", "v", false,
"verbose")
```

Now, rather than checking for every error if the `verbose` flag is called and printing out the error before it is returned, we create a simple function that can be repeated for checking but also returning the error value. Within the `utils/errors.go` file, we define the following function for reuse:

```
func Error(errString string, err error, verbose bool) error {
    errString = cleanup(errString, err)
    if err != nil {
        if verbose {
            // prints to stdout also
            Verbose.Error(errString)
        } else {
            Logger.Error(errString)
        }
        return fmt.Errorf(errString)
    }
    return nil
}
```

Let's take one command as an example, the `delete` command, which shows how this function is called:

```
var deleteCmd = &cobra.Command{
    Use: "delete",
    Short: "Delete audiofile by id",
    Long: `Delete audiofile by id. This command removes the
        entire folder containing all stored metadata`,
```

The bulk of the code for the command is usually found within the Run or RunE method, which receives the cmd variable, a *cobra.Command instance, and the args variable, which holds arguments within a slice of strings. Very early on, in each method, we create the client and extract any flags we might need—in this case, the verbose, silence, and id flags:

```
RunE: func(cmd *cobra.Command, args []string) error {
    client := &http.Client{
        Timeout: 15 * time.Second,
    }
    var err error
      silence, _ := cmd.Flags().GetBool("silence")
    verbose, _ := cmd.Flags().GetBool("verbose")
    id, _ := cmd.Flags().GetString("id")
    if id == "" {
        id, err = utils.AskForID()
        if err != nil {
            return utils.Error("\n %v\n try again and
                enter an id", err, verbose)
        }
    }
```

Next, we construct the request we are sending to the HTTP client, which uses the id value:

```
params := "id=" + url.QueryEscape(id)
path := fmt.Sprintf("http://%s:%d/delete?%s",
    viper.Get("cli.hostname"),
    viper.GetInt("cli.port"), params)
payload := &bytes.Buffer{}
req, err := http.NewRequest(http.MethodGet,
    path, payload)
if err != nil {
    return utils.Error("\n %v\n check configuration
        to ensure properly configured hostname and
        port", err, verbose)
}
```

We check whether there's any error when creating the request, which is most likely a result of a configuration error. Next, we log the request so that we are aware of any communication to external servers:

```
utils.LogRequest(verbose, http.MethodGet, path,
    payload.String())
```

We'll execute the request through the client's Do method and return an error if the request was unsuccessful:

```
resp, err := client.Do(req)
if err != nil {
    return utils.Error("\n %v\n check configuration
        to ensure properly configured hostname and
        port\n or check that api is running", err,
        verbose)
}
defer resp.Body.Close()
```

Following the request, we check the response and read the resp.Body, or the body of the response, if the response was successful. If not, an error message will be returned and logged:

```
err = utils.CheckResponse(resp)
if err != nil {
    return utils.Error("\n checking response: %v",
    err, verbose)
}
b, err := ioutil.ReadAll(resp.Body)
if err != nil {
    return utils.Error("\n reading response: %v
        \n ", err, verbose)
}
utils.LogHTTPResponse(verbose, resp, b)
```

Finally, we check whether the response returns the success string, which shows a successful deletion. The result is then printed out to the user:

```
if strings.Contains(string(b), "success") && !silence {
    fmt.Printf("\U00002705 Successfully deleted
        audiofile (%s)!\n", id)
} else {
```

```
                fmt.Printf("\U0000274C Unsuccessful delete of
                    audiofile (%s): %s\n", id, string(b))
            }
            return nil
        },
    }
```

You'll see that the utils.Error function is called every time an error is encountered. You'll also see a few other logging functions: utils.LogRequest and utils.LogHTTPResponse. The first, utils.LogRequest, is defined to log the request to either standard output, the log file, or both:

```
func LogRequest(verbose bool, method, path, payload string) {
    if verbose {
        Verbose.Info(fmt.Sprintf("sending request: %s %s
            %s...\n", method, path, payload))
    } else {
        Logger.Info(fmt.Sprintf("sending request: %s %s
            %s...\n", path, path, payload))
    }
}
```

The second, utils.LogHTTPResponse, similarly logs the response from the previous request to either standard output, the log file, or both:

```
func LogHTTPResponse(verbose bool, resp *http.Response, body []
byte) {
    if verbose && resp != nil {
        Verbose.Info(fmt.Sprintf("response status: %s,
            body: %s", resp.Status, string(body)))
    } else if resp != nil {
        Logger.Info(fmt.Sprintf("response status: %s, body:
            %s", resp.Status, string(body)))
    }
}
```

Now that this logger has been implemented for all the audiofile commands, let's give it a try and see what the output looks like now that the command has a verbose flag to output debug data when necessary.

Trying out verbose mode to view stack traces

After recompiling the project, we run the `delete` command with an invalid ID and pass the `verbose` command:

```
./bin/audiofile delete --id invalidID --verbose
{"level":"info","time":"2022-11-06 21:21:44","file":"utils/
logger.go:112","message":"sending request: GET http://
localhost:8000/delete?id=invalidID ...\n"}
{"level":"error","time":"2022-11-06 21:21:44","file":"utils/
errors.go:17","message":"checking response: \u001b[41;37mthe id
cannot be found\u001b[0m","stacktrace":"github.com/marianina8/
audiofile/utils.Error\n\t/Users/mmontagnino/Code/src/github.
com/marianina8/audiofile/utils/errors.go:17\ngithub.com/
marianina8/audiofile/cmd.glob..func2\n\t/Users/mmontagnino/
Code/src/github.com/marianina8/audiofile/cmd/delete.go:54\
ngithub.com/spf13/cobra.(*Command).execute\n\t/Users/
mmontagnino/Code/src/github.com/marianina8/audiofile/vendor/
github.com/spf13/cobra/command.go:872\ngithub.com/spf13/cobra.
(*Command).ExecuteC\n\t/Users/mmontagnino/Code/src/github.com/
marianina8/audiofile/vendor/github.com/spf13/cobra/command.
go:990\ngithub.com/spf13/cobra.(*Command).Execute\n\t/Users/
mmontagnino/Code/src/github.com/marianina8/audiofile/vendor/
github.com/spf13/cobra/command.go:918\ngithub.com/marianina8/
audiofile/cmd.Execute\n\t/Users/mmontagnino/Code/src/github.
com/marianina8/audiofile/cmd/root.go:21\nmain.main\n\t/Users/
mmontagnino/Code/src/github.com/marianina8/audiofile/main.
go:11\nruntime.main\n\t/usr/local/go/src/runtime/proc.go:250"}
Error: checking response: the id cannot be found
Usage:
  audiofile delete [flags]

Flags:
  -h, --help        help for delete
      --id string   audiofile id

Global Flags:
  -v, --verbose    verbose
```

Using the `verbose` flag, the debug statements are printed out, and when an error occurs, the stack trace is also output. This is important data for the user to share with the developer to debug what went wrong. Now, let us learn how to give the option to the user to submit a bug.

Effortless bug submission

Let us create a bug command using the Cobra generator for users to submit issues to the developers of the audiofile CLI:

```
cobra-cli add bug
bug created at /Users/mmontagnino/Code/src/github.com/
marianina8/audiofile
```

Now that we have the bug command created, the Run field is changed to extract details of the application and launch a web browser with the data already added and ready for the user to just finish off the submission with some extra details:

```
var bugCmd = &cobra.Command{
    Use: "bug",
    Short: "Submit a bug",
    Long: "Bug opens the default browser to start a bug
        report which will include useful system
        information.",
    RunE: func(cmd *cobra.Command, args []string) error {
        if len(args) > 0 {
            return fmt.Errorf("too many arguments")
        }
        var buf bytes.Buffer
        buf.WriteString(fmt.Sprintf("**Audiofile
            version**\n%s\n\n", utils.Version()))
        buf.WriteString(description)
        buf.WriteString(toReproduce)
        buf.WriteString(expectedBehavior)
        buf.WriteString(additionalDetails)

        body := buf.String()
        url := "https://github.com/marianina8/audiofile/issues/
new?title=Bug Report&body=" + url.QueryEscape(body)
        // we print if the browser fails to open
        if !openBrowser(url) {
            fmt.Print("Please file a new issue at https://
github.com/marianina8/audiofile/issues/new using this
template:\n\n")
```

```
        fmt.Print(body)
        }
        return nil
    },
}
```

The strings passed into the buf.WriteString method are defined outside the command within the same file, cmd/bug.go, but once the command is run, the complete template body is defined as follows:

```
**Audiofile version**
1.0.0

**Description**
A clear description of the bug encountered.

**To reproduce**
Steps to reproduce the bug.

**Expected behavior**
Expected behavior.

**Additional details**
Any other useful data to share.
```

Calling the ./bin/audiofile bug command launches the browser to open a new issue on the GitHub repo:

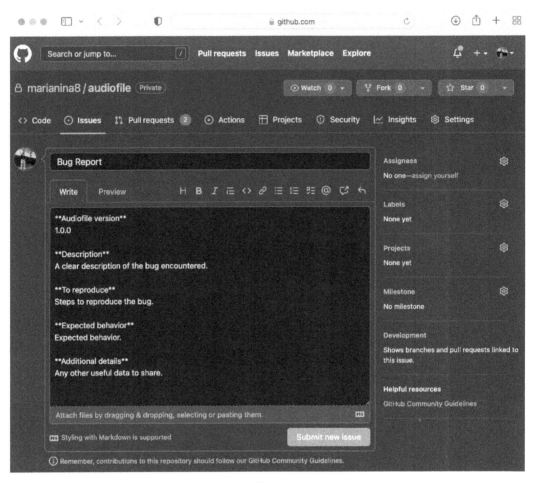

Figure 9.2 – Screenshot of browser open to a new issue

From the browser window, open the new issue page; the version of the CLI is populated, and then the user can replace the default text for the description, reproduction steps, expected behavior, and other steps with their own.

Help, documentation, and support

Part of creating a CLI that empathizes with its users is to supply sufficient help and documentation, as well as support users of all kinds. Luckily, the Cobra CLI framework supports the generation of help from the short and long fields of the Cobra command and the generation of man pages as well. However, bringing empathy into the extended documentation of your CLI may require several techniques.

Generating help text

By now, there have been many examples of creating commands, but just to reiterate, the command structure and the fields that show up in help are fields within the Cobra commands. Let's go over a good example:

```
var playCmd = &cobra.Command{
    Use: "play",
    Short: "Play audio file by id",
    Long: `Play audio file by id using the default audio
        player for your current system`,
    Example: `./bin/audiofile play -id
        45705eba-9342-4952-8cd4-baa2acc25188`,
    RunE: func(cmd *cobra.Command, args []string) error {
        // code
    },
}
```

Making sure you simply supply a short and long description of the command and one or several examples, you are supplying some help text that can at least get users started using the command. Running this will show the following output:

```
audiofile % ./bin/audiofile play --help
Play audio file by id using the default audio player for your
current system
Usage:
  audiofile play [flags]
Examples:
  ./bin/audiofile play -id 45705eba-9342-4952-8cd4-baa2acc25188
Flags:
  -h, --help        help for play
      --id string   audiofile id
Global Flags:
  -v, --verbose    verbose
```

A simple command doesn't need a ton of explanation, so this is enough to help guide the user with usage.

Generating man pages

In the `audiofile` repo, we've added some additional code to generate the man pages for the existing commands and commands in the `Makefile` to run to quickly run the code to do so. There exists a new program within the repo defined under `documentation/main.go`:

```
import (
    "log"

    "github.com/marianina8/audiofile/cmd"
    "github.com/spf13/cobra/doc"
)

func main() {
    header := &doc.GenManHeader{
        Title: "Audiofile",
        Source: "Auto generated by marianina8",
    }
    err := doc.GenManTree(cmd.RootCMD(), header, "./pages")
    if err != nil {
        log.Fatal(err)
    }
}
```

We pass in the `root` command and generate the pages in the `./pages` directory. The addition of the `make pages` command within the `Makefile` creates the man pages when called:

```
manpages:
    mkdir -p pages
    go run documentation/main.go
```

Within the terminal, if you run `make manpages` and then check to see whether the new pages exist by running `man pages/audiofile.1`, you will see the generated man page for the `audiofile` CLI:

```
● ○ ◉                          audiofile — less • man pages/audiofile.1 — 80×24
Audiofile(1)                                                          Audiofile(1)

NAME
       audiofile - A command line interface to interact with the Audiofile
       service

SYNOPSIS
       audiofile [flags]

DESCRIPTION
       A command line interface allows you to interact with the Audiofile
       service.  Basic commands include: get, list, and upload.

OPTIONS
       -h, --help[=false]  help for audiofile

       -t, --toggle[=false]    Help message for toggle

       -v, --verbose[=false]   verbose

SEE ALSO
       audiofile-api(1), audiofile-bug(1), audiofile-delete(1), audiofile-
       get(1), audiofile-list(1), audiofile-play(1), audiofile-search(1),
  :▮
```

Figure 9.3 – Screenshot of audiofile man pages in the terminal

You can also see that within the pages directory, there's an individual man page created for all the commands that have been added to the root command.

Embedding empathy into your documentation

By the time a user reaches your documentation, it is likely that they may already have encountered an issue and are frustrated or confused. It's important that your documentation takes in that perspective and portrays an understanding of the user's situation.

Although it may feel like documentation takes a lot of time and energy from other areas of development, it is essential for the future of your command-line application.

Within the last few years, there's been a recent term, *empathy advocacy*, that has come up in regard to technical documentation. It was coined by Ryan Macklin, a technical and UX writer, as well as an empathy advocate. The term is used to describe a subfield of technical communication centered on empathy and realistic respect for human emotion. It can be considered a framework for the way you communicate with your users. Because many people come to your documentation, we know that there's a varied assortment of brain chemistry, life experience, and recent events playing in mind. Empathy advocacy is one solution to this beautiful challenge.

Macklin has proposed seven philosophical documentation techniques rooted in empathy advocacy. These principles have been informed by disciplines such as UX, trauma psychotherapy, neurobiology, gameplay design, and cultural and language differences. Let's discuss each of these tenets and why they work:

- **Employ visual storytelling**—The human brain easily grabs onto stories, and sighted users can benefit from visuals. However, this forces developers to think about different types of accessibility: visual, cognitive, motor, and so on. Telling a story forces the writer to think about structure. On the other hand, dense and long-winded text is **accessibility-hostile**. As a note, this idea doesn't work for everyone.

- **Use synopses**—Using a **tl;dr** (short for **too long, don't read**), a summary line, or a banner provides a shortened explanation for tired and stressed-out readers who benefit from a lower cognitive cost option. Cognitive glue is required for running a collection of cognitive tasks to complete a high level of intelligence. Cognitive glue requires energy, so providing a synopsis will provide a low-cost option for users who are already running on low.

- **Give time frames**—In general, uncertainty creates **vicious voids**, and dwelling within the unknown time frame can create heightened emotional responses. Providing time frames can help stabilize the void. Time frames can be given if there's an outage on the server side, an upload to the server, or just a general time to complete a certain task.

- **Include short videos**—This is a great alternative for some users who struggle with reading comprehension. Typically, younger audiences are used to video, and when you split videos up into a single topic at max, the shorter playtime can be reassuring. Reassurance is a powerful way to regulate emotion. However, there are some pitfalls to video—mainly, that video costs more time and energy to create.

- **Reduce screenshots**—Providing screenshots can be helpful, but only when the UI can be confusing. Also, providing just enough for the user to figure some things out themselves helps to foster cognitive glue. Otherwise, being bombarded by visuals hurts everyone.

- **Rethink FAQs**—Instead of a traditional question and answer, break up documentation into single-scoped documents. Provide specific titles and avoid over-promising.

- **Pick your battles**—It's difficult to fight every fight; do the best you can, and choose your battles. Not everything you do will work for everyone—learn along the way. After all, advocating for empathy is another means of self-care.

Hopefully, these tenets that describe the philosophy of empathy advocacy help you to think twice about the words you use in your documentation. A few things to consider when you are writing your documentation include how your words may come across to someone who is in a panicked or frustrated state. Also, consider how you can help those about to give up or lacking the energy to complete their task to be successful.

Summary

In this chapter, you have learned specific steps to make your command-line application more empathetic. From error handling, debug and traceback information, effortless bug submission, and empathic advocacy in technical communication, you have learned the technical and empathic skills to apply within your application.

Errors can now be rewritten in color to jump out of the screen and decorated with additional information that provides the user information on exactly where an error has occurred and potentially what they may need to do to reach a resolution. When an error seems unresolvable, the user can then run the same command using the `--verbose` flag and view the detail logs, which might contain server requests and responses necessary to trace more specifically where an error may be happening, down to the line of code.

If a bug is encountered, the addition of a new `bug` command allows the user to spawn a new browser straight from their terminal, opening straight to a new template in GitHub's new issue submission form.

Finally, bridging the gap between technical documentation and the user's perspective is done by taking an empathetic approach. Several philosophical tenets when using an empathic framework when writing your documentation were discussed.

Questions

1. Which two common methods can you use for decorating your errors?

2. Between Zap and Logrus loggers, why would you choose Zap?

3. What is empathy advocacy?

Further reading

- *Empathy Advocacy*: `https://empathyadvocacy.org`

- *Write the Docs*: `https://www.writethedocs.org`

Answers

1. `fmt.Errorf(format string, a ...any) error` or `errors.Wrap(err error, message string) error`.

2. Zap is faster and is actively maintained.

3. Empathy advocacy is a sub-field of technical communication centered on empathy and realistic respect for human emotion. It can be considered a framework for the way you write your technical documentation and a solution for writing for many types of people with varied backgrounds and accessibilities.

10

Interactivity with Prompts and Terminal Dashboards

One powerful way to increase usability for users is to integrate interactivity with either prompts or terminal dashboards. Prompts are useful because they create a conversational approach while requesting input. Dashboards are useful because they allow developers to create a graphical interface from ASCII characters. That graphical interface, via a dashboard, can create powerful visual cues to allow users to navigate through different commands.

This chapter will give you examples of how to build user surveys from a series of prompts, and a terminal dashboard – whether learning about the Termdash library, designing the mockup, or implementing it for the audio file CLI.

Interactivity is fun. It's the more human and empathetic approach to a command-line interface. However, remember to disable interactivity if you are not outputting to a terminal. This chapter will cover the basics of surveys and dive deep into the terminal dashboard. By the end of this chapter, you'll have everything you need to create your own survey or dashboard. We will cover the following:

- Guiding users with prompts
- Designing a useful terminal dashboard
- Implementing a terminal dashboard

Technical requirements

- You'll need a Unix operating system to understand and run the examples shared in the chapter
- Get the `termdash` package at `https://github.com/mum4k/termdash`
- Get the survey package at `https://github.com/go-survey/survey`

- You can also find the code examples on GitHub at `https://github.com/PacktPublishing/Building-Modern-CLI-Applications-in-Go/tree/main/Chapter10`

Guiding users with prompts

There are many ways to simply prompt the user, but if you want to create a whole survey that can retrieve information using a variety of different prompts – text input, multi-select, single-select, multi-line text, password, and more – it might be useful to use a preexisting library to handle this for you. Let's create a generic customer survey using the `survey` package.

To show you how to use this package, I'll create a survey that can prompt the user for different types of input:

- **Text input** – for example, an email address
- **Select** – for example, a user's experience with the CLI
- **Multiselect** – for example, any issues encountered
- **Multiline** – for example, open-ended feedback

In the `Chapter-10` repository, a survey has been written to handle these four prompts. The questions, stored in the qs variable, are defined as a slice of `*survey.Question`:

```
questions := []*survey.Question{
    {
        Name: "email",
        Prompt: &survey.Input{
          Message: "What is your email address?"
    },
        Validate: survey.Required,
        Transform: survey.Title,
    },
    {
        Name: "rating",
        Prompt: &survey.Select{
            Message: "How would you rate your experience with
                      the CLI?",
            Options: []string{"Hated it", "Disliked", "Decent",
                              "Great", "Loved it"},
        },
```

```
        },
        {
            Name: "issues",
                Prompt: &survey.MultiSelect{
                Message: "Have you encountered any of these
                            issues?",
                Options: []string{"audio player issues", "upload
                                  issues", "search issues", "other
                                  technical issues"},
            },
        },
        {
            Name: "suggestions",
            Prompt: &survey.Multiline{
                Message: "Please provide any other feedback or
                            suggestions you may have.",
            },
        },
    }
```

We'll need an answers struct to store all the results from the prompts:

```
results := struct {
    Email string
    Rating string
    Issues []string
    Suggestions string
}{}
```

And finally, the method that asks the questions and stores the results:

```
err := survey.Ask(questions, &results)
if err != nil {
    fmt.Println(err.Error())
    return
}
```

Now that we've created the survey, we can try it out:

```
mmontagnino@Marians-MacBook-Pro Chapter-10 % go run main.go
? What is your email? mmontagnino@gmail.com
? How would you rate your experience with the CLI? Great
? Have you encountered any of these issues? audio player
issues, search issues
? Please provide any other feedback or suggestions you may
have. [Enter 2 empty lines to finish] I want this customer
survey embedded into the CLI and email myself the results!
```

Prompting the user is an easy way to integrate interactivity into your command-line application. However, there are even more colorful and fun ways to interact with your users. In the next section, we'll discuss the terminal dashboard, the `termdash` package in detail, and how to mock up and implement a terminal dashboard.

Designing a useful terminal dashboard

Command-line interfaces don't have to be limited to text. With **termdash**, a popular Golang library, you can build a terminal dashboard providing users with a user interface to visually see progress, alerts, text, and more. Colorful widgets placed within a clean dashboard that's been neatly laid out can increase information density and present a lot of information to the user in a very user-friendly manner. In this section, we'll learn about the library and the different layout choices and widget options. At the end of the chapter, we'll design a terminal dashboard that we can implement in our **audio file** command-line interface.

Learning about Termdash

Termdash is a Golang library that provides a customizable and cross-platform, terminal-based dashboard. On the project's GitHub page, a fun and colorful demo provides an example of all possible widgets demonstrated within a dynamic layout. From the demo, you can see that you can go all out on a fancy dashboard. To do so, you'll need to understand how to lay out a dashboard, interact with keyboard and mouse events, add widgets, and fine-tune the appearance with alignment and color. Within this section, we will break down the layers of a Termdash interface and the widgets that can be organized within it.

A Termdash dashboard consists of four main layers:

- The terminal layer
- The infrastructure layer

- The container layer
- The widgets layer

Let's take a deep dive into each of them.

The terminal layer

Think of the terminal layer of a dashboard as a 2D grid of cells that exist within a buffer. Each cell contains either an ASCII or Unicode character with the option to customize the foreground color, the color of text, the background color, or the color of the non-character space within the cell. Interactions with the mouse and keyboard happen on this layer as well.

Two terminal libraries can be used to interact at the cell level of a terminal:

- **tcell**: Inspired by **termbox** and has many new improvements
- **termbox**: No longer supported, although it is still an option

The following examples will utilize the `tcell` package to interact with the terminal. To start, create a new `tcell` instance to interact via the terminal API:

```
terminalLayer, err := tcell.New()
if err != nil {
    return err
}
defer terminalLayer.Close()
```

Notice that in this example, `tcell` has two methods: New and Close. New creates a new `tcell` instance in order to interact with the terminal and Close closes the terminal. It's a good practice to defer closing access to `tcell` right after creation. Although there are no options passed into the New method, there are a few optional methods that can be called:

- `ColorMode` sets the color mode when initializing a terminal
- `ClearStyle` sets the foreground and background color when a terminal is cleared

An example of initializing a cell in `ColorMode` to access all 256 available terminal colors would look like this:

```
terminalLayer, err := tcell.New(tcell.ColorMode(terminalapi.
ColorMode256))
if err != nil {
    return err
```

```
}
defer terminalLayer.Close()
```

ClearStyle, by default, will use ColorDefault if no specific ClearStyle is set. This ColorDefault is usually the default foreground and background colors of the terminal emulator, which are typically black and white. To set a terminal to use a yellow foreground and navy background style when the terminal is cleared, the New method, which accepts a slice of options, would be modified in the following way:

```
terminalLayer, err := tcell.New(tcell.ColorMode(terminalapi.
ColorMode256), tcell.ClearStyle(cell.ColorYellow, cell.
ColorNavy))
if err != nil {
    return err
}
defer terminalLayer.Close()
```

Now that we've created a new tcell that gives us access to the Terminal API, let's discuss the next layer – infrastructure.

The infrastructure layer

The infrastructure of a terminal dashboard provides the organization of the structure. The three main elements of the infrastructure layer include alignment, line style, and Termdash.

Alignment

Alignment is provided by the align package, which provides two alignment options – align. Horizonal, which includes predefined values of left, center, and right and align. Vertical with predefined values of top, middle, and bottom.

Line style

The line style defines the style of the line drawn on the terminal either when drawing boxes or borders.

The package exposes the options available via LineStyle. The LineStyle type represents a style that follows the Unicode options.

Termdash

Termdash provides the developer with the main entry point. Its most important purpose is to start and stop the dashboard application, control screen refreshing, process any runtime errors, and subscribe and listen for keyboard and mouse events. The termdash.Run method is the simplest way to start a Termdash application. The terminal may run until the context expires, a keyboard shortcut is called, or it times out. The simplest way to get started with the dashboard is with the following minimal code example, which creates a new `tcell` for the terminal layer, and a new **container** for the container layer. A container is another module within the `termdash` package, which we will dive into in the next section. We create context with a 2-minute timeout and then call the Run method of the `termdash` package:

```
if terminalLayer, err := tcell.New()
if err != nil {
    return err
}
defer terminalLayer.Close()

containerLayer, err := container.New(terminalLayer)
if err != nil {
    return err
}

ctx, cancel := context.WithTimeout(context.Background(),
1*time.Second)
defer cancel()
if err := termdash.Run(ctx, terminalLayer, containerLayer); err
!= nil {
    return err
}
```

In the preceding code example, the dashboard will run until the context expires, in 60 seconds.

Screen redrawing, or refreshing, for your Terminal dashboard can be done in a few ways: periodic, time-based redraws or manually triggered redraws. Only one method may be used, as using one means the other method is ignored. Besides that, the screen will refresh each time an input event occurs. The `termdash.RedrawInterval` method is an option that can be passed into the Run method to tell the dashboard application to redraw, or refresh, the screen at a particular interval. The Run method can be modified with the option to refresh every 5 seconds:

```
termdash.Run(ctx, terminalLayer, containerLayer, termdash.
RedrawInterval(5*time.Second))
```

The dashboard may also be redrawn using a controller, which can be triggered manually. This option means that the dashboard is drawn only once and unlike the Run method, the user maintains control of the main goroutine. An example of this code, using the previously defined `tcell` and `container` variables defined earlier, can be passed into a new controller to be drawn manually:

```
termController, err := termdash.NewController(terminalLayer,
containerLayer)
if err != nil {
    return err
}
defer termController.Close()


if err := termController.Redraw(); err != nil {
    return fmt.Errorf("error redrawing dashboard: %v", err)
}
```

The Termdash API provides a `termdash.ErrorHandler` option, which tells the dashboard how to handle errors gracefully. Without providing an implementation for this error handler, the dashboard will panic on all runtime errors. Errors can occur when processing or retrieving events, subscribing to an event, or when a container fails to draw itself.

An error handler is a callback method that receives an error and handles the error appropriately. It can be defined as a variable and, in the simplest case, just prints the runtime error:

```
errHandler := func(err error) {
    fmt.Printf("runtime error: %v", err)
}
```

When starting a Termdash application using the Run or NewController method, the error handler may be passed in as an option using the `termdash.ErrorHandler` method. For example, the Run method can be modified with a new option:

```
termdash.Run(ctx, terminalLayer, containerLayer, termdash.
ErrorHandler(errHandler))
```

While the NewController method can be modified similarly:

```
termdash.NewController(terminalLayer, containerLayer, termdash.
ErrorHandler(errHandler))
```

Through the `termdash` package, you can also subscribe to keyboard and mouse events. Typically, the container and certain widgets subscribe to keyboard and mouse events. Developers can also subscribe to certain mouse and keyboard events to take global action. For example, a developer may want the terminal to run a specific function when a specific key is set. `termdash.KeyboardSubscriber` is used to implement this functionality. With the following code, the user subscribes to the letters q and Q and responds to the keyboard events by running code to quit the dashboard:

```
keyboardSubscriber := func(k *terminalapi.Keyboard) {
    switch k.Key {
      case 'q':
      case 'Q':
          cancel()
    }
}

if err := termdash.Run(ctx, terminalLayer, containerLayer,
termdash.KeyboardSubscriber(keyboardSubscriber)); err != nil {
return fmt.Errorf("error running termdash with keyboard
subscriber: %v", err)
}
```

Another option is to call the Run method with the option to listen to mouse events using `termdash.MouseSubscriber`. Similarly, the following code can be called to do something random when the mouse button is clicked within the dashboard:

```
mouseClick := func(m *terminalapi.Mouse) {
    switch m.Button {
        case mouse.ButtonRight:
        // when the left mouse button is clicked - cancel
        cancel()
        case mouse.ButtonLeft:
        // when the left mouse button is clicked
        case mouse.ButtonMiddle:
        // when the middle mouse button is clicked
    }
}

if err := termdash.Run(ctx, terminalLayer, containerLayer,
termdash.MouseSubscriber(mouseClick)); err != nil {
```

```
        return fmt.Errorf("error running termdash with mouse
    subscriber: %v", err)
    }
```

The container layer

The container layer provides options for dashboard layouts, container styles, keyboard focus, and margin and padding. It also provides a method for placing a widget within a container.

From the previous examples, we see that a new container is called using the `container.New` function. We'll provide some new examples of how to organize your container and set it up with different layouts.

There are two main layout options:

- Binary tree
- Grid layouts

The **binary tree layout** organizes containers in a binary tree structure where each container is a node in a tree, which, unless empty, may contain either two sub-containers or a widget. Sub-containers can be split further with the same rules. There are two kinds of splits:

- **Horizontal splits**, created with the `container.SplitHorizontal` method, will create top and bottom sub-containers specified by `container.Top` and `container.Bottom`
- **Vertical splits**, created with the `container.SplitVertical` method, will create left and right sub-containers, specified by `container.Left` and `container.Right`

The `container.SplitPercent` option specifies the percentage of container split to use when spitting either vertically or horizontally. When the split percentage is not specified, a default of 50% is used. The following is a simple example of a binary tree layout using all the methods described:

```
    terminalLayer, err := tcell.New(tcell.
ColorMode(terminalapi.ColorMode256),
        tcell.ClearStyle(cell.ColorYellow, cell.ColorNavy))
    if err != nil {
        return fmt.Errorf("tcell.New => %v", err)
    }
    defer terminalLayer.Close()
leftContainer := container.Left(
container.Border(linestyle.Light),
)
rightContainer :=
```

```
container.Right(
container.SplitHorizontal(
container.Top(
container.Border(linestyle.Light),
),
container.Bottom(
container.SplitVertical(
        container.Left(
        container.Border(linestyle.Light),
        ),
        container.Right(
        container.Border(linestyle.Light),
        ),
        ),
          ),
      )
)
containerLayer, err := container.New(
terminalLayer,
container.SplitVertical(
leftContainer,
rightContainer,
container.SplitPercent(60),
),
)
```

Notice how we drill down when splitting up the terminal into containers. First, we split vertically to divide the terminal into left and right portions. Then, we split the right portion horizontally. The bottom-right horizontally split portion is split vertically. Running this code will present the following dashboard:

Figure 10.1 – Dashboard showing a container split using the binary layout

Notice that the container to the left takes up about 60% percent of the full width. The other splits do not define a percentage and take up 50% of the container.

The other option for a dashboard is to use a **grid layout**, which organizes the layout into rows and columns. Unlike the binary tree layout, the grid layout requires a grid builder object. Rows, columns, or widgets are then added to the grid builder object.

Columns are defined using either the `grid.ColWidhPerc` function, which defines a column with a specified width percentage of the parent's width, or `grid.ColWidthPercWithOpts`, which is an alternative that allows developers to additionally specify options when representing the column.

Rows are defined using either the `grid.RowHeightPerc` function, which defines a row with a specified height percentage of the parent's height, or `grid.RowHeightPercWithOpts`, which is an alternative that allows developers to additionally specify options when representing the row.

To add a widget within the grid layout, utilize the `grid.Widget` method. The following is a simple example of a layout implemented by the `grid` package. The code uses all the related methods and adds an ellipses text widget within each cell:

```
t, err := tcell.New()
if err != nil {
    return fmt.Errorf("error creating tcell: %v", err)
}
rollingText, err := text.New(text.RollContent())
if err != nil {
    return fmt.Errorf("error creating rolling text: %v",
      err)
}
```

```go
err = rollingText.Write("...")
if err != nil {
    return fmt.Errorf("error writing text: %v", err)
}
builder := grid.New()
builder.Add(
    grid.ColWidthPerc(60,
        grid.Widget(rollingText,
            container.Border(linestyle.Light),
        ),
    ),
)
builder.Add(
    grid.RowHeightPerc(50,
        grid.Widget(rollingText,
            container.Border(linestyle.Light),
        ),
    ),
)
builder.Add(
    grid.ColWidthPerc(20,
        grid.Widget(rollingText,
            container.Border(linestyle.Light),
        ),
    ),
)
builder.Add(
    grid.ColWidthPerc(20,
        grid.Widget(rollingText,
            container.Border(linestyle.Light),
        ),
    ),
)
gridOpts, err := builder.Build()
if err != nil {
    return fmt.Errorf("error creating builder: %v", err)
```

```
        }

        c, err := container.New(t, gridOpts...)
```

Running the code generates the following dashboard:

Figure 10.2 – Dashboard showing the container created using the grid layout

Notice that the column width percentage equals 100%; anything more would cause a compilation error.

There is also the option of a dynamic layout that allows you to switch between different layouts on the dashboard. Using the `container.ID` option, you can identify a container with some text, which can be referenced later so there's a way to identify which container will be dynamically updated using the `container.Update` method:

```
        t, err := tcell.New()
        if err != nil {
            return fmt.Errorf("error creating tcell: %v", err)
        }
        defer t.Close()

        b1, err := button.New("button1", func() error {
            return nil
        })
        if err != nil {
            return fmt.Errorf("error creating button: %v", err)
```

```go
}

b2, err := button.New("button2", func() error {
    return nil
})
if err != nil {
    return fmt.Errorf("error creating button: %v", err)
}

c, err := container.New(
    t,
    container.PlaceWidget(b1),
    container.ID("123"),
)
if err != nil {
    return fmt.Errorf("error creating container: %v", err)
}
update := func(k *terminalapi.Keyboard) {
    if k.Key == 'u' || k.Key == 'U' {
        c.Update(
            "123",
            container.SplitVertical(
                container.Left(
                    container.PlaceWidget(b1),
                ),
                container.Right(
                    container.PlaceWidget(b2),
                ),
            ),

        )
    }
}

ctx, cancel := context.WithTimeout(context.Background(),
    5*time.Second)
```

```
    defer cancel()
    if err := termdash.Run(ctx, t, c, termdash.
        KeyboardSubscriber(update)); err != nil {
            return fmt.Errorf("error running termdash: %v", err)
    }
```

In this code, the container ID is set to 123. Originally, the widget contained just one button. The update method replaces the single button with a container split vertically, with one button on the left and another on the right. When running this code, pressing the *u* key runs the update on the layout.

The original layout shows a single button:

Figure 10.3 – Layout showing a single button

After pressing the *u* or *U* key, the layout updates:

Figure 10.4 – Layout showing two buttons after pressing the u key again

The container layer can be further configured using margin and padding settings. The margin is the space outside of the container's border while the padding is the space between the inside of the container's border and its content. The following image provides the best visual representation of margins and padding:

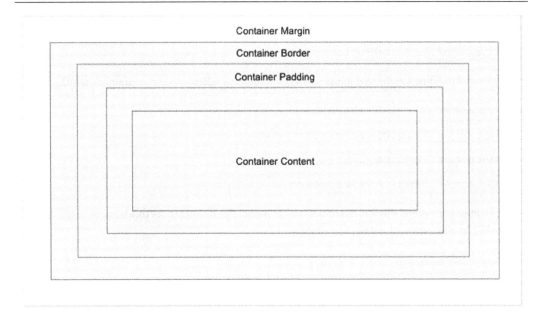

Figure 10.5 – Margin and padding

The margin and padding can be set with either absolute or relative values. An absolute margin can be set with the following options:

- `container.MarginTop`
- `container.MarginRight`
- `container.MarginBottom`
- `container.MarginLeft`

Absolute padding can be set with the following options:

- `container.PaddingTop`
- `container.PaddingRight`
- `container.PaddingBottom`
- `container.PaddingLeft`

Relative values for the margin and padding are set with percentages. The margin and padding's top and bottom percentage values are relative to the container's height:

- `container.MarginTopPercent`
- `container.MarginBottomPercent`

- container.PaddingTopPercent
- container.PaddingBottomPercent

The margin and padding's right and left percentage values are relative to the container's width:

- container.MarginRightPercent
- container.MarginLeftPercent
- container.PaddingRightPercent
- container.PaddingLeftPercent

Another form of placement within containers is alignment. The following methods are available from the align API to align content within the container:

- container.AlignHorizontal
- container.AlignVertical

Let's put it all together in a simple example that extends upon the binary tree code example:

```
b, err := button.New("click me", func() error {
    return nil
})
if err != nil {
    return err
}
leftContainer :=
container.Left(
    container.Border(linestyle.Light),
        container.PlaceWidget(b),
        container.AlignHorizontal(align.HorizontalLeft),
    )
rightContainer :=
        container.Right(
            container.SplitHorizontal(
                container.Top(
                    container.Border(linestyle.Light),
                    container.PlaceWidget(b),
                    container.AlignVertical(align.VerticalTop),
                ),
```

```
                    container.Bottom(
                      container.SplitVertical(
                          container.Left(
                            container.Border(linestyle.Light),
                                container.PlaceWidget(b),
                                container.PaddingTop(3),
                                container.PaddingBottom(3),
                                container.PaddingRight(3),
                                container.PaddingLeft(3),
                          ),
                          container.Right(
                            container.Border(linestyle.Light),
                                container.PlaceWidget(b),
                                container.MarginTop(3),
                                container.MarginBottom(3),
                                container.MarginRight(3),
                                container.MarginLeft(3),
                          ),
                      ),
                  ),
              ),
                  )
containerLayer, err := container.New(
        terminalLayer,
        container.SplitVertical(
            leftContainer,
            rightContainer,
            container.SplitPercent(60),
        ),
    )
```

The resulting layout appears as follows:

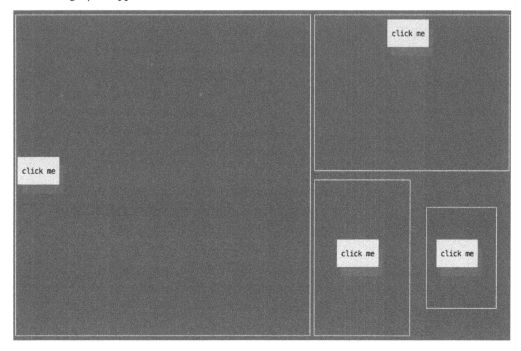

Figure 10.6 – Container showing different alignments for a button, with different margins and padding

You can also define a key to change the focus to the next or previous container using the `container.KeyFocusNext` and `container.KeyFocusPrevious` options.

The widget layer

In several of the previous examples, we showed code that placed a widget in either a grid or binary tree container layout and also customized the alignment, margin, and padding. However, besides a simple button or text, there are different widget options, and the demo on the GitHub page shows an example of each:

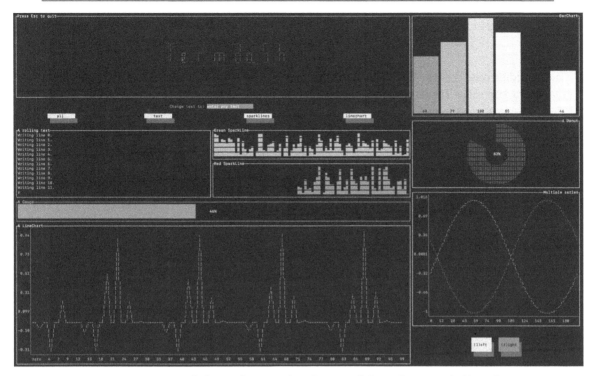

Figure 10.7 – Termdash sample screenshot showing all the widgets in a dashboard

Let's do a quick example of each with a snippet of code to understand how each widget is created. To add each widget to a container, just use the `container.PlaceWidget` method that was used earlier for the simple text and button examples. Let's go over a few other examples: a bar chart, donut, and gauge. For a detailed code of the other widgets, visit the very well-documented termdash wiki and check out the demo pages.

A bar chart

Here is some example code for creating a bar chart widget with individual values displayed relative to a max value:

```
barChart, err := barchart.New()
if err != nil {
    return err
}
values := []int{20, 40, 60, 80, 100}
max := 100
if err := barChart.Values(values, max); err != nil {
```

```
        return err
    }
```

The preceding code creates a new `barchart` instance and adds the values, a slice of `int`, plus the maximum `int` value. The resulting terminal dashboard looks like this:

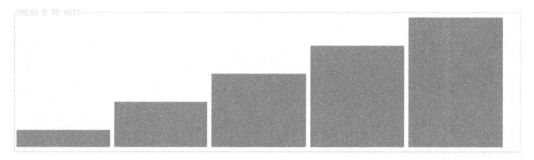

Figure 10.8 – Bar chart example

Change the values of the `values` and `max` variables to see the chart change. The color of the bars can also be modified based on preference.

A donut

A donut, or progress circle chart, represents the completion of progress. Here is some example code for creating a donut chart to show percentages:

```
greenDonut, err := donut.New(
    donut.CellOpts(cell.FgColor(cell.ColorGreen)),
    donut.Label("Green", cell.FgColor(cell.ColorGreen)),
)
if err != nil {
    return err
}
greenDonut.Percent(75)
```

The preceding code creates a new `donut` instance with options for the label and foreground color set to green. The resulting terminal dashboard looks like this:

Figure 10.9 – Green donut at 75%

Again, the color can be modified based on preference, and remember, since Termdash provides dynamic refreshing, the data can be automatically updated and redrawn, making it quite nice for showing progress.

A gauge

A gauge, or progress bar, is another way to measure the amount completed. The following is some sample code for showing how to create a progress gauge:

```
progressGauge, err := gauge.New(
    gauge.Height(1),
    gauge.Border(linestyle.Light),
    gauge.BorderTitle("Percentage progress"),
)
if err != nil {
    return err
}
progressGauge.Percent(75)
```

This code creates a new instance of a gauge with options for a light border, a title, **Percentage progress**, and a slim height of 1. The percentage, as with the donut, is 75%. The resulting terminal dashboard looks like this:

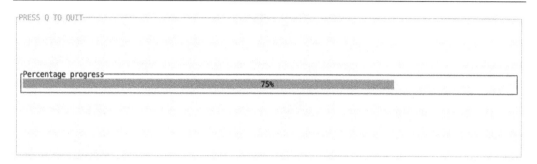

Figure 10.10 – Gauge at 75% percent progress

As mentioned before, because of dynamic redrawing, this is another great option for showing progress updates.

Now that we've shown examples of different widgets to include within a terminal dashboard, let's sketch out a design using these widgets that we can later implement in our audio file command-line interface. Suppose we wanted to build a music player in a terminal dashboard. Here is a sample layout:

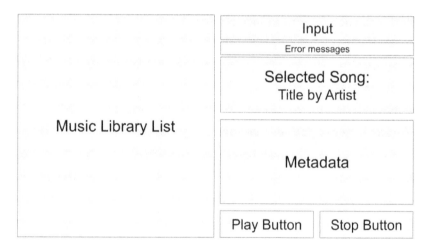

Figure 10.11 – Terminal dashboard layout

This layout can be created easily with the binary layout. The music library list section can be generated from a list of songs with number identifiers, which can be used in the text input section, where a song can be selected by ID. Any error messages associated with the input ID will be displayed right below. If the input is good, the selected song section will show rolling ASCII text with the song title, and the metadata section will display the text metadata of the selected song. Hitting the play button will start playing the selected song, and the stop button will stop it. Proceed to the next section where we'll make this terminal dashboard a reality.

Implementing a terminal dashboard

When creating a terminal dashboard, you can create it as a separate standalone application or as a command that is called from the command-line application. In our specific example for the player terminal dashboard, we are going to call the dashboard when the `./bin/audiofile player` command is called.

First, from the audio file's root repository, we'll need to use `cobra-cli` to create the command:

```
cobra-cli add player
Player created at /Users/mmontagnino/Code/src/github.com/
marianina8/audiofile
```

Now, we can create the code to generate the terminal dashboard, called within the Run field of the `player` command. Remember that the terminal dashboard consists of four main layers: the terminal, infrastructure, container, and widgets. Like a painting, we'll start with the base layer: the terminal.

Creating the terminal layer

The first thing you need to do is to create a terminal that provides access to any input and output. Termdash has a `tcell` package for creating a new `tcell`-based terminal. Many terminals by default only support 16 colors, but other more modern terminals can support up to 256 colors. The following code specifically creates a new terminal with a 265-color mode.

```
t, err := tcell.New(tcell.ColorMode(terminalapi.ColorMode256))
```

After creating a terminal layer, we then create the infrastructure layer.

Creating the infrastructure layer

The infrastructure layer handles the terminal setup, mouse and keyboard events, and containers. In our terminal dashboard player, we want to handle a few tasks:

- Keyboard event to signal quitting
- Running the terminal dashboard, which subscribes to this keyboard event

Let's write the code to handle these two features required of the terminal dashboard.

Subscribing to keyboard events

If we want to listen for key events, we create a keyboard subscriber to specify the keys to listen to:

```
quitter := func(k *terminalapi.Keyboard) {
    if k.Key == 'q' || k.Key == 'Q' {
```

```
        . . .
    }
}
```

Now that we have defined a keyboard subscriber, we can use this as an input parameter to termdash's Run method.

Running the terminal

When running the terminal, you'll need the terminal variable, container, and keyboard and mouse subscribers, as well as the timed redrawing interval and other options. The following code runs the tcell-based terminal we created and the quitter keyboard subscriber, which listens for *q* or *Q* key events to quit the application:

```
if err := termdash.Run(ctx, t, c, termdash.
KeyboardSubscriber(quitter), termdash.RedrawInterval(100*time.
Millisecond)); err != nil {
    panic(err)
}
```

The c variable that's passed into the termdash.Run method as the third parameter is the container. Let's define the container now.

Creating the container layer

When creating the container, it helps to look at the bigger picture of the layout and then narrow it down as you go. For example, when you first look at the planned layout, you'll see the largest sections are made from left and right vertical splits.

Figure 10.12 – Initial vertical split

As we begin to define the container, we'll slowly drill down with more specifics, but we begin with the following:

- **Vertical Split (Left)** – The music library
- **Vertical Split (Right)** – All other widget

The final code reflects this drill-down process. Since we keep the left vertical split as the music library, we drill down with containers on the left, always starting with the larger containers and adding smaller ones within.

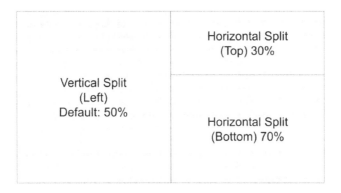

Figure 10.13 – Horizontal split of right vertical space

The next is a horizontal split that separates the left vertical split into the following:

- **Horizontal Split (Top) 30%** – Text input, error messages, and the rolling song title text
- **Horizontal Split (Bottom) 70%** – Metadata and play/stop buttons

Let's take the top horizontal split and split it, again, horizontally:

- **Horizontal Split (Top) 30%** – Text input and error message
- **Horizontal Split (Bottom) 70%** – The rolling song title text

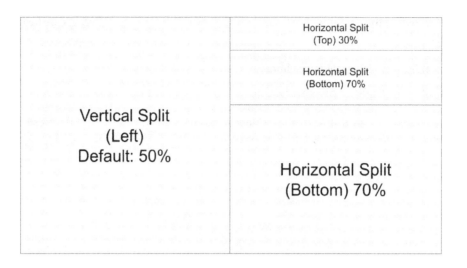

Figure 10.14 – Horizontal split of top horizontal space

We split the earlier top part horizontally into the separated text input and error messages:

- **Horizontal Split (Top) 60%** – Text input
- **Horizontal Split (Bottom) 40%** – Error messages

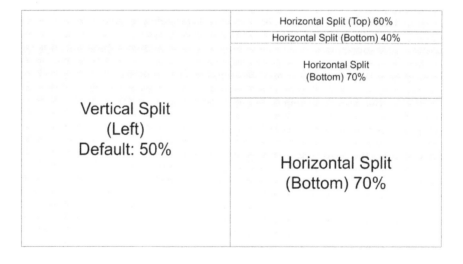

Figure 10.15 – Horizontal split of top horizontal space

Now, let's drill down into the bottom 70% of the initial horizontal split of the right vertical container. Let's split it up into two horizontal sections:

- **Horizontal Split (Top) 80%** – The metadata section
- **Horizontal Split (Bottom) 20%** – The button section (play/stop)

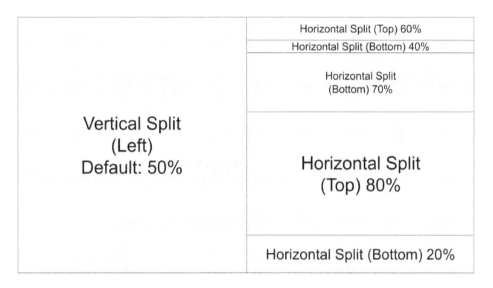

Figure 10.16 – Horizontal split of bottom horizontal space

Finally, the last part to drill down to is the bottom horizontal split, which we will split vertically:

- **Vertical Split (Left) 50%** – The play button
- **Vertical Split (Right) 50%** – The stop button

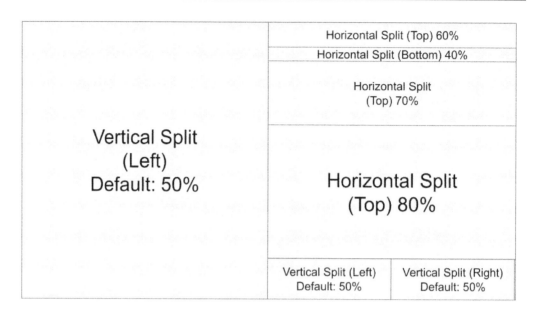

Figure 10.17 – Vertical split of bottom horizontal space

The entire layout broken down with the container code shows this drill-down process – I've added comments for where the widgets will be placed for reference:

```
c, err := container.New(
    t,
    container.SplitVertical(
        container.Left(), // music library
        container.Right(
            container.SplitHorizontal(
                container.Top(
                    container.SplitHorizontal(
                        container.Top(
                            container.SplitHorizontal(
                                container.Top(), // text input
                                container.Bottom(), // error
                                                    msgs
                                container.SplitPercent(60),
                            ),
                        ),
```

```
                    container.Bottom(), // rolling song
                                               title
                    container.SplitPercent(30),
                ),
            ),
            container.Bottom(
                container.SplitHorizontal(
                    container.Top(), // metadata
                    container.Bottom(
                        container.SplitVertical(
                            container.Left(), // play
                                               button
                            container.Right(), // stop
                                               button
                        )
                    ),
                    container.SplitPercent(80),
                ),
            ),
            container.SplitPercent(30),
        ),
    ),
)
```

Next, let's create the widgets and place them within the appropriate containers to finalize the terminal dashboard.

Creating the widgets layer

Going back to the original layout, all the different widgets we'll need to implement are clear to see:

- The music library list
- Input text
- Error messages
- Rolling text – selected song (title by artist)
- Metadata

- The play button

- The stop button

At this point, I am aware of which widget to use for each item on the list. However, if you have not yet decided, now is the time to determine the best Termdash widget to use for each item:

- Text:

  - Music library list

  - Error messages

  - Rolling text – selected song (title by artist), metadata

- Text input:

  - Input field

- Button:

  - The play button

  - The stop button

Let's create at least one of each type as an example. The full code is available in the `Chapter10` GitHub repository.

Creating a text widget for the music library list

The music library list will take in the audio list and print the text in a section that will list the index of the song next to the title and artist. We define this widget with the following function:

```
func newLibraryContent(audioList *models.AudioList) (*text.
Text, error) {
    libraryContent, err := text.New(text.RollContent(), text.
      WrapAtWords())
    if err != nil {
        panic(err)
    }
    for i, audiofile := range *audioList {
        libraryContent.Write(fmt.Sprintf("[id=%d] %s by %s\n",
          i, audiofile.Metadata.Tags.Title, audiofile.Metadata.
          Tags.Artist))
    }
```

```
        return libraryContent, nil
    }
```

The function is called in the Run function field like so:

```
    libraryContent, err := newLibraryContent(audioList)
```

The error message and metadata items are also text widgets, so we'll omit those code examples. Next, we'll create the input text.

Creating an input text widget for setting the current ID of a song

The input text section is where a user inputs the ID of the song displayed in the music library section. The input text is defined within the following function:

```
func newTextInput(audioList *models.AudioList, updatedID chan<-
int, updateText, errorText chan<- string) *textinput.TextInput
{
    input, _ := textinput.New(
        textinput.Label("Enter id of song: ", cell.
          FgColor(cell.ColorNumber(33))),
        textinput.MaxWidthCells(20),
        textinput.OnSubmit(func(text string) error {
            // set the id
            // set any error text
        return nil
    }),
    textinput.ClearOnSubmit(),
    )
    return input
}
```

Creating a button to start playing the song associated with the input ID

The last type of widget is a button. There are two different buttons we need, but the following code is for the play button:

```
func newPlayButton(audioList *models.AudioList, playID <-chan
int) (*button.Button, error) {
    playButton, err := button.New("Play", func() error {
        stopTheMusic()
```

```
        }
        go func() {
        if audiofileID <= len(*audioList)-1 && audiofileID >= 0
{
        pID, _ = play((*audioList)[audiofileID].Path, false,
                true)
        }}()
        return nil
    },
    button.FillColor(cell.ColorNumber(220)),
    button.GlobalKey('p'),
    )
    if err != nil {
        return playButton, fmt.Errorf("%v", err)
    }
    return playButton, nil
}
```

- The function is called in the Run function field:

```
playButton, err := newPlayButton(audioList, playID)
```

- Once all the widgets have been created, they are placed within the container in the appropriate places with the following line of code:

```
container.PlaceWidget(widget)
```

- Once the widgets have been placed within the container, we can run the terminal dashboard with the following command:

```
./bin/audiofile player
```

- Magically, the player terminal dashboards appear and we can select an ID to enter and play a song:

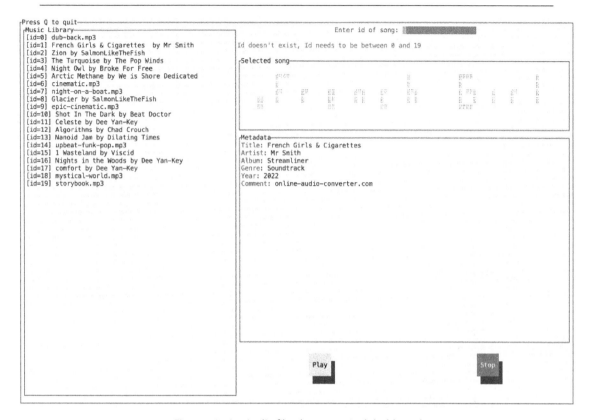

Figure 10.18 – Audio file player terminal dashboard

- Voila! We've created a terminal dashboard to play the music in our audio file library. While you can view the metadata through the command-line application's get and list commands and play music with the play command, the new player terminal dashboard allows you to view what exists in the audio file library in a more user-friendly fashion.

Summary

In this chapter, you learned how to create a survey with different interactive prompts and a terminal dashboard containing a variety of widgets. These are just examples that can hopefully inspire you in terms of interactivity within your own command-line application.

The survey example showed you how to use a variety of different types of prompts; you can prompt the user for their user experience, but as you've seen within the audio file CLI, you can also just prompt for missing information. These prompts can be input throughout your code in places where prompts may come in handy, or they can be strung along a list of other questions and you can create a more thorough survey for your users.

The player terminal dashboard gives you an example of how to create a terminal dashboard for a command-line interface. Consider the kind of data your users will be sending or retrieving from your command-line interface and let that guide you in your design of a more visual approach.

Questions

1. What method is used to create the terminal layer?

2. What method is used to place a widget inside a container?

3. What's the difference between the binary layout and the grid layout?

Answers

1. `tcell.New()`

2. `container.PlaceWidget(widget)`

3. The grid layout allows you to split the container into horizontal rows and vertical columns. The binary layout allows you to split sub-containers horizontally or vertically.

Further reading

- *The Big Book of Dashboards: Visualizing Your Data Using Real-World Business Scenarios* by Wexler, Shaffer, and Cotgreave

Part 4:
Building and Distributing
for Different Platforms

This part of the book is all about building, testing, and distributing your CLI application using Docker and GoReleaser. It starts by explaining the importance of building and testing, and how build tags with Boolean logic can be used to create targeted builds and testing to further stabilize your project with each new feature. The part also covers cross-compilation, a powerful feature of Go, which enables you to compile your application for different operating systems and architectures. The benefits of containerization are also explored, with a focus on Docker containers for testing and distributing your apps. Finally, we end with a discussion using GoReleaser and GitHub Actions in tandem to automate the release of a CLI application as a Homebrew formula, which makes it easy for MacOS users to find and install your software with just one command.

This part has the following chapters:

- *Chapter 11, Custom Builds and Testing CLI Commands*
- *Chapter 12, Cross Compilation Across Different Platforms*
- *Chapter 13, Using Containers for Distribution*
- *Chapter 14, Publishing your Go binary as a Homebrew Formula with GoReleaser*

11
Custom Builds and Testing CLI Commands

With any Golang application, you'll need to build and test. However, it is increasingly important as the project and its user base grow. Build tags with Boolean logic give you the ability to create targeted builds and testing and further stabilize your project with each new feature.

Given a deeper understanding of build tags and how to use them, we will use a real-world example, the audio file CLI, to integrate levels (free and pro) and enable a profiling feature.

Build tags are not only used as input when building but also when testing. We will spend the latter half of this chapter on testing. We will learn specifically how to mock an HTTP client that our CLI is using, configure tests locally, write tests for individual commands, and run them. In this chapter, we will cover the following topics in detail:

- What are build tags and how can you use them?
- Building with tags
- Testing CLI commands

Technical requirements

- You'll need a Unix operating system to understand and run the examples shared in the chapter
- You can also find the code examples on GitHub at `https://github.com/PacktPublishing/Building-Modern-CLI-Applications-in-Go/tree/main/Chapter11/audiofile`

What are build tags and how can you use them?

Build tags are indicators of when a code file should be included within the build process. In Go, they are defined by a single line at the top, or near the top, of any source file, not just a Go file. They must precede the package clause and be followed by a blank line. They have the following syntax:

```
//go:build [tag]
```

This line can only be defined once in a file. More than one definition would generate an error. However, when more than one tag is used, they interact using Boolean logic. In *Chapter 7*, *Developing for Different Platforms*, we briefly touched on tags and their logic. The other method for handling the development of different platforms uses a series of `if-else` statements that check the operating system at runtime. Another method is to include the operating system in the filename. For example, if there's a filename ending in `_windows.go`, we indicate to the compiler to only include this file when building for `windows`.

Tags can help separate code to include when compiling for different operating systems using `$GOOS` and `$GOARCH`. Valid combinations of operating systems and the architecture can be found here: `https://go.dev/doc/install/source#environment`.

Besides targeting platforms, build tags can be customized to separate featured code or integration tests. Often, integration tags receive a specific tag, as they often take a longer time to run. Separating unit tests from integration tests adds a level of control when testing your application.

These build constraints, when used together, can powerfully compile different versions of your code. As mentioned, they are evaluated together using Boolean logic. Expressions contain build tags combined using the `||`, `&&`, and `!` operators and parentheses. To learn more about build constraints, run the following command in your terminal:

```
go help buildconstraint
```

As an example, the following build tags constrain a file to build when the `linux` or `openbsd` tags are satisfied and when `amd64` is satisfied and `cgo` is not:

```
//go:build (linux  || openbsd) && amd64 && !cgo
```

Run `go env` in your terminal to see which tags are satisfied automatically when building your application. You'll see the target operating system (`$GOOS`) and architecture (`$GOARCH`) and `unix` if the operating system is Unix or Unix-like. The `cgo` field is determined by the `CGO_ENABLED` environment variable, the term for each Go major release, and any additional tags given by the `-tags` flag.

As mentioned earlier, you can create your own pro and free versions based on tags placed at the top of code files, `//go:build pro` or `//go:build free`. Integration test files can be tagged with `//go:build int`, for example. However you want to customize your builds, you can do so with the power of tags and Boolean logic. Now, in the next section, let's use tags in our code to do just that.

How to utilize build tags

As mentioned, we can use build tags to separate builds based on the operating system and architecture. Within the audio file repository, we're already doing so with the following files associated with the `play` and `bug` commands. For the `bug` command, we have the following files:

- `bug_darwin.go` // only builds on Darwin systems

- `bug_linux.go` // only builds on Linux systems

- `bug_windows.go` // only builds on Windows platforms

Each of those files contains a function that is specifically coded for the targeted platform. The file suffixes have similar functionality to the build tags. You can choose a file suffix that matches the exact platform and architecture. However, build tags are preferred when you want to target more than one platform and architecture. Inside the files is the matching build tag, used as an example, but duplicates functionality. Inside `bug_darwin.go`, for example, at the top of the file is the following:

```
//go:build darwin
```

Since we already have these build tags set up throughout the repo to target platforms where needed, let's explore a few other ways to utilize build tags.

Creating a pro, free, and dev version

Suppose the command-line interface utilized build tags to create different levels of access to the application's features. This could be for admin or basic level users or restricted by the level of permissions, but it could also be, especially if the CLI was for external customers, a pro and free level version of your application.

First, it's important to decide which commands will be available for each version. Let's give this a try with the audio file application:

Command	Free	Pro
bug	x	x
delete	x	x
get	x	x
list	x	x
play	x	x
player		x
search		x
upload	x	x

Table 11.1 – List of commands included in the free or pro level

Let's also include a dev version; this simply allows the API to be run locally. In a real-world scenario, the application would be configured to call a public API, and storage could be done in a database. This gives us another build tag to create.

Now, let's use build tags to distinguish the free, pro, and dev versions. The dev version build tag is placed at the top of the cmd/api.go file, making the API command only available when the dev tag is specified:

```
//go:build dev
```

Then, the tag to distinguish the pro version is as follows:

```
//go:build !free && pro
```

There are a few files, as previously mentioned, that already have build tags to target platforms. This build tag means that the file will be available in the free, pro, and dev versions:

```
//go:build darwin
```

The preceding build tags utilize Boolean logic to state that the file should be included in the build process when both the darwin and free tags are defined.

Let's break down the tags here with the Boolean logic syntax examples:

Build Tag Syntax	Build Tag Sample	Boolean Statement
|| separated	//go:build (free || pro)	free OR pro
&& separated	//go:build darwin && free	darwin AND free
! exclamation point	//go:build !free	NOT free

Table 11.2 – Boolean logic examples

This Boolean logic included within the build tag will allow developers to build for any combination of platforms and versions.

Adding build tags to enable pprof

Another way to utilize build tags is to enable profiling on your API service. `pprof` is a tool for visualizing and analyzing profile data. The tool reads a collection of samples in `proto`, or protocol buffer, format and then creates reports that help visualize and analyze the data. This tool can generate text and graphical reports.

> **Note**
> To learn more about how to use this tool, visit `https://pkg.go.dev/net/http/pprof`.

For this case, we'll define a build tag called `pprof` to appropriately match its usage. Within the `services/metadata/metadata.go` file, we define the metadata service used to extract information from the audio files uploaded via the command-line interface. The `CreateMetadataService` function creates the metadata service and defines all the endpoints with matching handlers. To enable profiling, we will add this new block of code:

```
if profile {
    mux.HandleFunc("/debug/pprof/", pprof.Index)
    mux.HandleFunc("/debug/pprof/{action}", pprof.Index)
    mux.HandleFunc("/debug/pprof/symbol", pprof.Symbol)
}
```

At the top of the file, after the inputs, we'll define the variable that it's dependent on:

```
var (
    profile = false
)
```

However, we need some way to set the `profile` variable to `true`. To do so, we create a new file: `services/metadata/pprof.go`. This file contains the following content:

```
//go:build profile && (free || pro)

package metadata

func init() {
    profile = true
}
```

As you can see, whether building the `free`, `pro`, or `dev` version, if the `profile` build tag is added as tag input, then the `init` function will be called to set the `profile` variable to `true`. Now, we have another idea of how to use build tags – to set Boolean variables that act as feature flags. Now that we've changed the necessary files to include the build tags, let's use these as inputs to the build commands.

Building with tags

By now, we have built our applications using `Makefile`, which contains the following command specific to building a Darwin application:

```
build-darwin:
    go build -tags darwin -o bin/audiofile main.go
    chmod +x bin/audiofile
```

For the Darwin build, we can additionally build a version for a free and pro version and also a profile version to enable `pprof`.

Building a free version

To build a `free` version for the Darwin operating system, we need to modify the preceding `make` command and create a new one:

```
build-darwin-free:
    go build -tags "darwin free" -o bin/audiofile main.go
    chmod +x bin/audiofile
```

In the build-darwin-free command, we pass in the two build tags: darwin and free. This will include files such as bug_darwin.go and play_darwin.go, which contain the following line at the top of the Go file:

```
//go:build darwin
```

Similarly, the files will be included in the build when we build the pro version.

Building a pro version

To build a pro version for the Darwin operating system, we need to add a new build command:

```
build-darwin-pro:
    go build -tags "darwin pro" -o bin/audiofile main.go
    chmod +x bin/audiofile
```

In the build-darwin-pro command, we pass in the two build tags: darwin and pro.

Building to enable pprof on the pro version

To build a pro version that has pprof enabled, we add the following build command:

```
build-darwin-pro-profile:
    go build -tags "darwin pro profile" -o bin/audiofile main.
go
    chmod +x bin/audiofile
```

In the build-darwin-pro-profile command, we pass three build tags: darwin, pro, and profile. This will include the services/metadata/pprof.go file, which includes the line at the top of the file:

```
//go:build profile
```

Similarly, the files will be included in the build when we build for the free version.

At this point, we've learned what build tags are, the different ways to use build tags within your code, and, finally, how to build applications targeted to specific uses using build tags. Specifically, while build tags can be used to define different levels of features available (free versus pro), you can also enable profiling or any other debug tooling using build tags. Now that we have understood how to build our command-line application for different targets, let's learn how to test our CLI commands.

Testing CLI commands

While building your command-line application, it's important to also build testing around it so you can ensure that the application works as expected. There are a few things that typically need to be done, including the following:

1. Mock the HTTP client
2. Handle test configuration
3. Create a test for each command

We'll go over the code for each of these steps that exist in the audio file repository for *Chapter 11*.

Mocking the HTTP client

To mock the HTTP client, we'll need to create an interface to mimic the client's Do method, as well as a function that returns this interface, which is both satisfied by the real and mocked client.

In the cmd/client.go file, we've written some code to handle all of this:

```
type AudiofileClient interface {
    Do(req *http.Request) (*http.Response, error)
}

var (
    getClient = GetHTTPClient()
)

func GetHTTPClient() AudiofileClient {
    return &http.Client{
        Timeout: 15 * time.Second,
    }
}
```

We can now easily create a mock client by replacing the getClient variable with a function that returns a mocked client. If you look at each command's code, it uses the getClient variable. For example, the upload.go file calls the Do method with the following line:

```
resp, err := getClient.Do(req)
```

When the application runs, this returns the actual HTTP client with a 15-second timeout. However, in each test, we'll set the getClient variable to a mocked HTTP client.

The mocked HTTP client is set in the `cmd/client_test.go` file. First, we define the type:

```
type ClientMock struct {
}
```

Then, to satisfy the `AudiofileClient` interface previously defined, we implement the Do method:

```
func (c *ClientMock) Do(req *http.Request) (*http.Response,
error) {
```

Some of the requests, including `list`, `get`, and `search` endpoints, will return data that is stored in JSON files under the `cmd/testfiles` folder. We read these files and store them in the corresponding byte slices: `listBytes`, `getBytes`, and `searchBytes`:

```
listBytes, err := os.ReadFile("./testfiles/list.json")
if err != nil {
    return nil, fmt.Errorf("unable to read testfile/list.json")
}
getBytes, err := os.ReadFile("./testfiles/get.json")
if err != nil {
    return nil, fmt.Errorf("unable to read testfile/get.json")
}
searchBytes, err := os.ReadFile("./testfiles/search.json")
if err != nil {
    return nil, fmt.Errorf("unable to read testfile/search.
json")
}
```

The data read from these files is used within the response. Since the Do method receives the request, we can create a switch case for each request endpoint and then handle the response individually. You can create more detailed cases to handle errors, but in this case, we are only returning the successful case. For the first case, the `/request` endpoint, we return `200 OK`, but the body of the response also contains the string value from `getBytes`. You can see the actual data in the `./testfiles/get.json` file:

```
switch req.URL.Path {
    case "/request":
        return &http.Response{
            Status:  "OK",
            StatusCode: http.StatusOK,
```

```
            Body: ioutil.NopCloser(bytes.
    NewBufferString(string(getBytes))),
        ContentLength: int64(len(getBytes)),
        Request: req,
        Header: make(http.Header, 0),
    }, nil
```

For the /upload endpoint, we return 200 OK, but the body of the response also contains the "123" string value:

```
        case "/upload":
            return &http.Response{
                Status:  "OK",
                StatusCode: http.StatusOK,
        Body: ioutil.NopCloser(bytes.NewBufferString("123")),
        ContentLength: int64(len("123")),
        Request: req,
        Header: make(http.Header, 0),
    }, nil
```

For the /list endpoint, we return 200 OK, but the body of the response also contains the string value from listBytes. You can see the actual data in the ./testfiles/list.json file:

```
        case "/list":
            return &http.Response{
                Status:  "OK",
                StatusCode: http.StatusOK,
                Body: ioutil.NopCloser(bytes.
                    NewBufferString(string(listBytes))),
                    ContentLength: int64(len(listBytes)),
                    Request: req,
                    Header: make(http.Header, 0),
    }, nil
```

For the /delete endpoint, we return 200 OK, but the body of the response also contains "successfully deleted audio with id: 456":

```
        case "/delete":
            return &http.Response{
                Status:  "OK",
```

```
                        StatusCode: http.StatusOK,
                        Body: ioutil.NopCloser(bytes.
                            NewBufferString("successfully deleted
                               audio with id: 456")),
                            ContentLength: int64(len("successfully
                                         deleted audio with id:
                                         456")),
                        Request: req,
                        Header: make(http.Header, 0),
        }, nil
```

For the /search endpoint, we return 200 OK, but the body of the response also contains the string value from searchBytes. You can see the actual data in the ./testfiles/search.json file:

```
        case "/search":
            return &http.Response{
                Status:  "OK",
                StatusCode: http.StatusOK,
                Body: ioutil.NopCloser(bytes.
                NewBufferString(string(searchBytes))),
                ContentLength: int64(len(list searchBytes
                Bytes)),
                Request: req,
                Header: make(http.Header, 0),
        }, nil
        }
        return &http.Response{}, nil
        }
```

Finally, if the request path doesn't match any of the endpoints in the switch statement, then an empty response is returned.

Handling test configuration

We handle the test configuration in the cmd/root_test.go file:

```
var Logger *zap.Logger
var Verbose *zap.Logger


func ConfigureTest() {
```

```
    getClient = &ClientMock{}
    viper.SetDefault("cli.hostname", "testHostname")
    viper.SetDefault("cli.port", 8000)
    utils.InitCLILogger()
}
```

Within the `ConfigureTest` function, we set the `getClient` variable to a pointer to the `ClientMock` type. Because the `viper` configuration values are checked when the command is called, we set some default values for the CLI's hostname and port to random test values. Finally, in this file, the regular logger, `Logger`, and verbose logger, `Verbose`, are both defined and then later initialized by the `utils.InitCLILogger()` method call.

Creating a test for a command

Now that we have the mocked client, configuration, and loggers set up, let's create a test for the commands. Before I dive into the code for each, it's important to mention the line of code that's reused at the start of each test:

```
ConfigureTest()
```

The preceding section discusses the details of this function, but it prepares each state with a mocked client, default configuration values, and initialized loggers. In our examples, we use the `testing` package, which provides support for automated tests in Go. It is designed to be used in concert with the `go test` command, which executes any function in your code defined with the following format:

```
func TestXxx(*testing.T)
```

Xxx can be replaced with anything else, but the first character needs to be capital. The name itself is used to identify the type of test that is being executed. I won't go into each individual test, just three as examples. To view the entirety of tests, visit the audio file repository for this chapter.

Testing the bug command

The function for testing the `bug` command is defined here. It takes a single parameter, which is a pointer to the `testing.T` type, and fits the function format defined in the last section. Let's break down the code:

```
func TestBug(t *testing.T) {
    ConfigureTest()
    b := bytes.NewBufferString("")
    rootCmd.SetOut(b)
    rootCmd.SetArgs([]string{"bug", "unexpected"})
```

```
err := rootCmd.Execute()
if err != nil {
    fmt.Println("err: ", err)
}
actualBytes, err := ioutil.ReadAll(b)
if err != nil {
    t.Fatal(err)
}

expectedBytes, err := os.ReadFile("./testfiles/bug.txt")
if err != nil {
    t.Fatal(err)
}
if strings.TrimSpace(string(actualBytes)) != strings.
    TrimSpace(string(expectedBytes)) {
        t.Fatal(string(actualBytes), "!=",
            string(expectedBytes))
    }
}
```

In this function, we first define the output buffer, b, which we can later read for comparison to the expected output. We set the arguments using the SetArgs method and pass in an unexpected argument. The command is executed with the rootCmd.Execute() method and the actual result is read from the buffer and saved in the actualBytes variable. The expected output is stored within the ./testfiles/bug.txt file and is read into the expectedBytes variable. We compare these values to ensure that they are equal. Since we passed in an unexpected argument, the command usage is printed out. This test is designed to pass; however, if the trimmed strings are not equal, the test fails.

Testing the get command

The function for testing the get command is defined here. Similarly, the function definition fits the format to be picked up in the go test command. Remember the mocked client and that the get command calls the /request endpoint. The response body contains the value found in the ./testfiles/get.json file. Let's break down the code:

```
func TestGet(t *testing.T) {
    ConfigureTest()
    b := bytes.NewBufferString("")
    rootCmd.SetOut(b)
```

We pass in the following arguments to mimic the `audiofile get -id 123 -json` call:

```
rootCmd.SetArgs([]string{"get", "--id", "123", "--json"})
```

We execute the root command with the preceding arguments:

```
err := rootCmd.Execute()
if err != nil {
    fmt.Println("err: ", err)
}
```

We read the actual data output from `rootCmd`'s execution and store it in the `actualBytes` variable:

```
actualBytes, err := ioutil.ReadAll(b)
if err != nil {
    t.Fatal(err)
}
```

We read the expected data output from the `./testfiles/get.json` file:

```
expectedBytes, err := os.ReadFile("./testfiles/get.json")
if err != nil {
    t.Fatal(err)
}
```

Then, the data of both `actualBytes` and `expectedBytes` is unmarshalled into the `models. Audio` struct and then compared:

```
var audio1, audio2 models.Audio
json.Unmarshal(actualBytes, &audio1)
json.Unmarshal(expectedBytes, &audio2)
if !(audio1.Id == audio2.Id &&
audio1.Metadata.Tags.Album == audio2.Metadata.Tags.Album &&
audio1.Metadata.Tags.AlbumArtist == audio2.Metadata.Tags.
AlbumArtist &&
audio1.Metadata.Tags.Artist == audio2.Metadata.Tags.Artist
&&
audio1.Metadata.Tags.Comment == audio2.Metadata.Tags.
Comment &&
audio1.Metadata.Tags.Composer == audio2.Metadata.Tags.
Composer &&
```

```
    audio1.Metadata.Tags.Genre == audio2.Metadata.Tags.Genre &&
    audio1.Metadata.Tags.Lyrics == audio2.Metadata.Tags.Lyrics
&&
    audio1.Metadata.Tags.Year == audio2.Metadata.Tags.Year) {
        t.Fatalf("expected %q got %q", string(expectedBytes),
string(actualBytes))
    }
}
```

This test was designed to succeed, but if the data is not as expected, then the test fails.

Testing the upload command

The function for testing the upload command is defined here. Again, the function definition fits the format to be picked up in the go test command. Remember the mocked client and that the upload command calls the /upload endpoint with a mocked response body containing the "123" value. Let's break down the code:

```
func TestUpload(t *testing.T) {
    ConfigureTest()
    b := bytes.NewBufferString("")
    rootCmd.SetOut(b)
    rootCmd.SetArgs([]string{"upload", "--filename", "list.
                go"})
    err := rootCmd.Execute()
    if err != nil {
        fmt.Println("err: ", err)
    }
    expected := "123"
    actualBytes, err := ioutil.ReadAll(b)
    if err != nil {
        t.Fatal(err)
    }
    actual := string(actualBytes)
    if !(actual == expected) {
        t.Fatalf("expected \"%s\" got \"%s\"", expected,
                actual)
    }
}
```

`rootCmd`'s arguments are set to mimic the following command call:

```
audiofile upload -filename list.go
```

The file type and data are not validated because that happens on the API side, which is mocked. However, since we know the body of the response contains the `123` value, we set the expected variable to `123`. The `actual` value, which contains the output of the command execution, is then later compared to the expected one. The test is designed for success, but if the values are not equal, then the test fails.

We've now gone over several examples of how to test a CLI Cobra command. You can now create your own tests for your CLI, by mocking your own HTTP client and creating tests for each individual command. We haven't done so in this chapter, but it's good to know that build tags can also be used to separate different kinds of tests – for example, integration tests and unit tests.

Running the tests

To test your commands, you can run `go test` and pass in a few additional flags:

- `-v` for verbose mode
- `-tags` for any files you want to specifically target

In our test, we want to target just the `pro` build tag because that will cover all commands. We add two additional `Makefile` commands, one to run tests in verbose mode and one that doesn't:

```
test:
  go test ./... -tags pro
test-verbose:
  go test -v ./... -tags pro
```

After saving the `Makefile` from the terminal, you can execute the command:

```
make test
```

The following output is expected:

```
go test ./cmd -tags pro
ok      github.com/marianina8/audiofile/cmd
```

We now know how to run the tests utilizing build tags as well. This should be all the tools needed to run your own CLI testing.

Summary

In this chapter, you learned what build tags are and how to use them for different purposes. Build tags can be used for generating builds of different levels, separating our specific tests, or adding debug features. You also learned how to generate builds with the build tags that you added to the top of your files and how to utilize the Boolean logic of tags to quickly determine whether files will or won't be included.

You also learned how to test your Cobra CLI commands with Golang's default `testing` package. Some necessary tools were also included, such as learning how to mock an HTTP client. Together with the build tags, you can now not only build targeted applications with tags but also run tests with the same tags to target specific tests. In the next chapter, *Chapter 12, Cross-Compilation Across Different Platforms*, we will learn how to use these tags and compile for the different major operating systems: `darwin`, `linux`, and `windows`.

Questions

1. Where does the build tag go in a Golang file and what is the syntax?

2. What flag is used for both `go build` and `go test` to pass in the build tags?

3. What build tag could you place on an integration test Golang file and how would you run `go test` with the tag?

Answers

1. It's placed at the top of the file, before the package declaration, followed by a single empty line. The syntax is: `//go:build [tag]`.

2. The `-tags` flag is used to pass in build tags for both the `go build` and `go test` methods.

3. You could add the `//go:build int` build tag at the top of any integration test file, and then modify the test file to run this command: `go test ./cmd -tags "pro int"`.

Further reading

*   Read more about the `build` package at `https://pkg.go.dev/go/build`, and read more about the `testing` package at `https://pkg.go.dev/testing`

12

Cross-Compilation across Different Platforms

This chapter introduces the user to cross-compilation, a powerful feature of Go, across different platforms. While build automation tools exist, understanding how to cross-compile provides essential knowledge for debugging and customization when necessary. This chapter will explain the different operating systems and architectures that Go can compile and how to determine which is needed. After Go is installed in your environment, there is a command, go env, with which you can see all the Go-related environment variables. We will discuss the two major ones used for building: GOOS and GOARCH.

We will give examples of how to build or install an application for each major operating system: Linux, macOS, and Windows. You will learn how to determine the Go operating system and architecture settings based on your environment and the available architectures for each major operating system.

This chapter ends with an example script to automate cross-compilation across the major operating systems and architectures. A script to run on the Darwin, Linux, or Windows environments is provided. In this chapter, we will cover the following topics in detail:

- Manual compilation versus build automation tools
- Using GOOS and GOARCH
- Compiling for Linux, macOS, and Windows
- Scripting to compile for multiple platforms

Technical requirements

- You can also find the code examples on GitHub at https://github.com/ PacktPublishing/Building-Modern-CLI-Applications-in-Go/tree/ main/Chapter12/audiofile

Manual compilation versus build automation tools

In *Chapter 14, Publishing Your Go Binary as a Homebrew Formula with GoReleaser*, we will delve into a fantastic open source tool, **GoReleaser**, which automates the process of building and releasing Go binaries. Despite its power and usefulness, it's crucial to know how to manually compile your Go code. You see, not all projects can be built and released with GoReleaser. For instance, if your application requires unique build flags or dependencies, manual compilation may be necessary. Moreover, understanding how to manually compile your code is essential for addressing issues that may crop up during the build process. In essence, tools such as GoReleaser can make the process a lot smoother, but having a good grasp of the manual compile process is vital to ensure that your **command-line interface (CLI)** applications can be built and released in various scenarios.

Using GOOS and GOARCH

When developing your command-line application, it is important to maximize the audience by developing for as many platforms as possible. However, you may also want to target just a particular set of operating systems and architectures. In the past, it was much more difficult to deploy to platforms that differed from the one you were developing on. In fact, developing on a macOS platform and deploying it on a Windows machine involved setting up a Windows build machine to build the binary. The tooling would have to be synchronized, and there would be other deliberations that made collaborative testing and distribution cumbersome.

Luckily, Golang has solved this by building support for multiple platforms directly into the language's toolchain. As discussed in *Chapter 7, Developing for Different Platforms*, and *Chapter 11, Custom Builds and Testing CLI Commands*, we learned how to write platform-independent code and use the `go build` command and build tags to target specific operating systems and architectures. You may also use environment variables to target the operating system and architecture as well.

First, it's good to know which operating systems and architectures are available for distribution. To find out, within your terminal, run the following command:

```
go tool dist list
```

The list is output in the following format: GOOS/GOARCH. GOOS is a local environment variable that defines the operating system to compile for and stands for **Go Operating System**. GOARCH, pronounced "gore-ch," is a local environment variable that defines the architecture to compile for and stands for **Go Architecture**.

```
aix/ppc64        linux/386       netbsd/arm64
android/386      linux/amd64     openbsd/386
android/amd64    linux/arm       openbsd/amd64
android/arm      linux/arm64     openbsd/arm
android/arm64    linux/loong64   openbsd/arm64
darwin/amd64     linux/mips      openbsd/mips64
darwin/arm64     linux/mips64    plan9/386
dragonfly/amd64  linux/mips64le  plan9/amd64
freebsd/386      linux/mipsle    plan9/arm
freebsd/amd64    linux/ppc64     solaris/amd64
freebsd/arm      linux/ppc64le   windows/386
freebsd/arm64    linux/riscv64   windows/amd64
illumos/amd64    linux/s390x     windows/arm
ios/amd64        netbsd/386      windows/arm64
ios/arm64        netbsd/amd64
js/wasm          netbsd/arm
```

Figure 12.1 – List of supported operating systems and architectures

You can also call the preceding command with the −json flag to view more details. For example, for linux/arm64, you can see that it's supported by Cgo from the "CgoSupported" field, but also that it is a first-class **port**, another word for a GOOS/GOARCH pair, indicated by the "FirstClass" field:

```
{
"GOOS": "linux",
"GOARCH": "arm64",
"CgoSupported": true,
"FirstClass": true
},
```

A first-class port has the following properties:

- Releases are blocked by broken builds
- Official binaries are provided
- Installation is documented

Next, determine your local operating system and architecture settings by running the following command within your terminal:

```
go env GOOS GOARCH
```

Currently, running this command on my macOS machine with an AMD64 architecture gives the following output:

```
darwin
amd64
```

The first environment variable, GOOS, is set to darwin, and the second environment variable, GOARCH, is set to amd64. We now know what GOOS and GOARCH are within the Go environment, the possible values, and also what values are set on your machine. Let's learn how to use these environment variables.

You can use these two environment variables for compiling. Let's generate a build to target the darwin/amd64 port. You'll do so by setting the GOOS or GOARCH environment variables and then running the go build command, or more specifically along with the build command:

```
GOOS=darwin GOARCH=amd64 go build
```

Let's try this out with the audio file CLI and learn all the ways to compile for the three main operating systems: Linux, macOS, and Windows.

Compiling for Linux, macOS, and Windows

There are several different ways to compile our command-line application for different operating systems and we'll go over examples of each of these. First, you can compile by building or installing your application:

- **Building** – Compiles the executable file and then moves it to the current folder or the filename indicated by the -o (output) flag
- **Installing** – Compiles the executable file and then installs it to the $GOPATH/bin folder or $GOBIN if it is set and caches all non-main packages, which are imported to the $GOPATH/pkg folder

Building using tags

In our previous chapter, *Chapter 11, Custom Builds and Testing CLI Commands*, we learned to build specifically for the macOS or Darwin operating system. To better understand how to use the build command, we run go build -help to see the usage:

```
mmontagnino@Marians-MacBook-Pro audiofile % go build -help
usage: go build [-o output] [build flags] [packages]
Run 'go help build' for details
```

Running `go help build` will reveal the build flags available. However, in these examples, we only use the `tags` flag. Within the `Makefile`, we already have the following commands:

```
build-darwin-free:
    go build -tags "darwin free" -o bin/audiofile main.go
    chmod +x bin/audiofile

build-darwin-pro:
    go build -tags "darwin pro" -o bin/audiofile main.go
    chmod +x bin/audiofile

build-darwin-pro-profile:
    go build -tags "darwin pro profile" -o bin/audiofile main.
go
    chmod +x bin/audiofile
```

In these commands, we compile the application and output it to the `bin/audiofile` filename. To specify the Darwin operating system, we pass in the Darwin build tag to specify the files associated with the Darwin operating system. We'll need to modify the output files to a folder that specifies Darwin, but also for other specifics such as the free versus the pro version since we'll be building for other operating systems and levels. Let's modify these.

Building applications for a Darwin operating system using tags

The new `Makefile` commands to compile the application for the Darwin operating system are now as follows:

```
build-darwin-free:
    go build -tags "darwin free" -o builds/free/darwin/
audiofile main.go
    chmod +x builds/free/darwin/audiofile

build-darwin-pro:
    go build -tags "darwin pro" -o builds/pro/darwin/audiofile
main.go
    chmod +x builds/pro/darwin/audiofile

build-darwin-pro-profile:
```

```
    go build -tags "darwin pro profile" -o builds/profile/
darwin/audiofile main.go
    chmod +x builds/profile/darwin/audiofile
```

We've swapped out the `bin/audiofile` output to something more specific. The free version for Darwin now outputs to `builds/free/darwin/audiofile`, the pro version outputs to `builds/pro/darwin/audiofile`, and the profile version outputs to `builds/profile/darwin/audiofile`. Let's continue with the next operating system, Linux.

We can do the same for Linux and Windows, like so:

```
build-linux-free:
    go build -tags "linux free" -o builds/free/linux/audiofile
main.go
    chmod +x builds/free/linux/audiofile


build-linux-pro:
    go build -tags "linux pro" -o builds/pro/linux/audiofile
main.go
    chmod +x builds/pro/linux/audiofile


build-linux-pro-profile:
    go build -tags "linux pro profile" -o builds/profile/linux/
audiofile main.go
    chmod +x builds/profile/linux/audiofile


build-windows-free:
    go build -tags "windows free" -o builds/free/windows/
audiofile.exe main.go


build-windows-pro:
    go build -tags "windows pro" -o builds/pro/windows/
audiofile.exe main.go


build-windows-pro-profile:
    go build -tags "windows pro profile" -o builds/profile/
windows/audiofile.exe main.go
```

The free Windows version is output to `builds/free/windows/audiofile.exe`, the pro Windows version is output to `builds/pro/windows/audiofile.exe`, and the Windows profile version is output to `builds/profile/windows/audiofile.exe`. Now, suppose we don't want to run each of the individual commands one by one, as there are so many to run! We can write a command to build all versions using tags.

Building applications for all operating systems using tags

Let's add a new `Makefile` command to build all the operating systems. Basically, we write one command that calls all other commands:

```
build-all: build-darwin-free build-darwin-pro build-darwin-
pro-profile build-linux-free build-linux-pro build-linux-pro-
profile build-windows-free build-windows-pro build-windows-pro-
profile
```

Let's try running this command via the terminal:

```
make build-all
```

If you're running on Darwin, you'll see the following output:

```
mmontagnino@Marians-MacBook-Pro audiofile % make build-all
go build -tags "darwin free" -o builds/free/darwin/audiofile
main.go
chmod +x builds/free/darwin/audiofile
go build -tags "darwin pro" -o builds/pro/darwin/audiofile
main.go
chmod +x builds/pro/darwin/audiofile
go build -tags "darwin pro profile" -o builds/profile/darwin/
audiofile main.go
chmod +x builds/profile/darwin/audiofile
go build -tags "linux free" -o builds/free/linux/audiofile
main.go
# internal/goos
/usr/local/go/src/internal/goos/zgoos_linux.go:7:7: GOOS
redeclared in this block
        /usr/local/go/src/internal/goos/zgoos_darwin.go:7:7:
other declaration of GOOS
/usr/local/go/src/internal/goos/zgoos_linux.go:9:7: IsAix
redeclared in this block
        /usr/local/go/src/internal/goos/zgoos_darwin.go:9:7:
other declaration of IsAix
```

```
/usr/local/go/src/internal/goos/zgoos_linux.go:10:7: IsAndroid
redeclared in this block
...
/usr/local/go/src/internal/goos/zgoos_linux.go:17:7: too many
errors
make: *** [build-linux-free] Error 2
```

I've removed part of the error message; however, the most important message is GOOS redeclared in this block. This error message comes up when the operating system is set but conflicts with the GOOS environment variable. For example, the command that failed used the operating build tag to specify a Linux build:

```
go build -tags "linux free" -o builds/free/linux/audiofile
main.go
```

However, running go env | grep GOOS in my macOS terminal shows the value of the GOOS environment variable:

```
GOOS="darwin"
```

Let's modify the build commands to set the GOOS environment variable so it matches the output type based on the build tag.

Building using the GOOS environment variable

The Linux builds have been modified to set the GOOS environment variable to Linux by prepending GOOS=linux before the build command:

```
build-linux-free:
    GOOS=linux go build -tags "linux free" -o builds/free/
linux/audiofile main.go
    chmod +x builds/free/linux/audiofile


build-linux-pro:
    GOOS=linux go build -tags "linux pro" -o builds/pro/linux/
audiofile main.go
    chmod +x builds/pro/linux/audiofile


build-linux-pro-profile:
```

```
    GOOS=linux go build -tags "linux pro profile" -o builds/
profile/linux/audiofile main.go
    chmod +x builds/profile/linux/audiofile
```

The Windows builds have been modified to set the GOOS environment variable to Windows by prepending GOOS=windows before the build command:

```
build-windows-free:
    GOOS=windows go build -tags "windows free" -o builds/free/
windows/audiofile.exe main.go


build-windows-pro:
    GOOS=windows go build -tags "windows pro" -o builds/pro/
windows/audiofile.exe main.go


build-windows-pro-profile:
    GOOS=windows go build -tags "windows pro profile" -o
builds/profile/windows/audiofile.exe main.go
```

Now, let's try the build-all command again. It runs successfully and we can see all the files generated by the build command by running find -type -f ./builds in the repo:

```
mmontagnino@Marians-MacBook-Pro audiofile % find ./builds -type
f
./builds/pro/linux/audiofile
./builds/pro/darwin/audiofile
./builds/pro/windows/audiofile.exe
./builds/free/linux/audiofile
./builds/free/darwin/audiofile
./builds/free/windows/audiofile.exe
./builds/profile/linux/audiofile
./builds/profile/darwin/audiofile
./builds/profile/windows/audiofile.exe
```

Building using the GOARCH environment variable

Many different possible architecture values can be associated with a single operating system. Rather than creating a command for each, we'll start with just one example:

```
build-darwin-amd64-free:
    GOOS=darwin GOARCH=amd64 go build -tags "darwin free" -o
builds/free/darwin/audiofile main.go
    chmod +x builds/free/darwin/audiofile
```

This example specifies the operating system, the GOOS environment variable, as darwin, and then the architecture, the GOARCH environment variable, as amd64.

There'd be too many commands to create if we were to create a build command for each architecture of each major operating system. We'll save this for a script within the last section of this chapter.

Installing using tags and GOOS env va

- As mentioned earlier, another way to compile your command-line application is by installing it. The install command compiles the application, like the go build command, but also with the additional step of moving the compiled application to the $GOPATH/bin folder or $GOBIN value. To learn more about the install command, we run the following go install -help command:

```
mmontagnino@Marians-MacBook-Pro audiofile % go install -help
usage: go install [build flags] [packages]
Run 'go help install' for details
```

- The same flags for building are available for installing. Again, we will use the tags flag only. Let's first run the install command on the macOS system:

```
go install -tags "darwin pro" github.com/marianina8/audiofile
```

However, running go env | grep GOPATH in my macOS terminal shows the value of the GOOS environment variable:

```
mmontagnino@Marians-MacBook-Pro audiofile % go env | grep
GOPATH
GOPATH="/Users/mmontagnino/Code"
```

Confirm that the audio file CLI executable exists in the $GOPATH/bin or /Users/mmontagnino/Code/bin folder.

As mentioned, we can use build tags to separate builds based on the operating system and architecture. Within the audio file repository, we're already doing so with the following files associated with the `play` and `bug` commands. For the `bug` command, we have the following files. Now, let's add some `install` commands within the `Makefile` now that we understand how to use the build tags and GOOS environment variables.

install commands for the Darwin operating system

The `install` commands for the Darwin operating system include passing in the specific tags, including `darwin`, and the levels, defined by tags, to install:

```
install-darwin-free:
    go install -tags "darwin free" github.com/marianina8/
audiofile

install-darwin-pro:
    go install -tags "darwin pro" github.com/marianina8/
audiofile

install-darwin-pro-profile:
    go install -tags "darwin pro profile" github.com/
marianina8/audiofile
```

install commands for the Linux operating system

The `install` commands for the Linux operating system include passing in the specific tags, including `linux`, and the package to install. To ensure the commands do not error out with conflicting GOOS settings, we set the matching environment variable, GOOS, to `linux`:

```
install-linux-free:
    GOOS=linux go install -tags "linux free" github.com/
marianina8/audiofile

install-linux-pro:
    GOOS=linux go install -tags "linux pro" github.com/
marianina8/audiofile

install-linux-pro-profile:
    GOOS=linux go install -tags "linux pro profile" github.com/
marianina8/audiofile
```

install commands for the Windows operating system

The `install` commands for the Windows operating system include passing in the specific tags, including `windows`, and the package to install. To ensure the commands do not error out with conflicting GOOS settings, we set the matching environment variable, GOOS, to `windows`:

```
install-windows-free:
    GOOS=windows go install -tags "windows free" github.com/
marianina8/audiofile
```

```
install-windows-pro:
    GOOS=windows go install -tags "windows pro" github.com/
marianina8/audiofile
```

```
install-windows-pro-profile:
    GOOS=windows go install -tags "windows pro profile" github.
com/marianina8/audiofile
```

Remember that for your `Makefile`, you'll need to change the location of the package if you have forked the repo under your own account. Run the `make` command for the operating system you need and confirm that the application is installed by checking the `$GOPATH/bin` or `$GOBIN` folder.

Installing using tags and GOARCH env var

While many different possible architecture values can be associated with a single operating system, let's start with just one example of installing with GOARCH `env var`:

```
install-linux-amd64-free:
    GOOS=linux GOARCH=amd64 go install -tags "linux free"
github.com/marianina8/audiofile
```

This example specifies the operating system, the GOOS environment variable, as `linux`, and then the architecture, the GOARCH environment variable, as `amd64`. Rather than creating a command for each pair of operating systems and architectures, again, we'll save this for a script within the last section of this chapter.

Scripting to compile for multiple platforms

We've learned several different ways to compile for operating systems using the GOOS and GOARCH environment variables and using build tags. The Makefile can fill up rather quickly with all the different combinations of GOOS/GOARCH pairs and scripting may provide a better solution if you want to generate builds for many more specific architectures.

Creating a bash script to compile in Darwin or Linux

Let's start by creating a bash script. Let's name it build.sh. To create the file, I simply type the following:

```
touch build.sh
```

The preceding command creates the file when it does not exist. The file extension is .sh, which, while unnecessary to add, clearly indicates that the file is a bash script type. Next, we want to edit it. If using vi, use the following command:

```
vi build.sh
```

Otherwise, edit the file using the editor of your choice.

Adding the shebang

The first line of a bash script is called the **shebang**. It is a character sequence that indicates the program loader's first instruction. It defines which interpreter to run when reading, or interpreting, the script. The first line to indicate to use the bash interpreter is as follows:

```
#!/bin/bash
```

The shebang consists of a couple of elements:

- #! instructs the program loader to load an interpreter for the code
- /bin/bash indicates the bash or interpreter's location

These are some typical shebangs for different interpreters:

Interpreter	Shebang
Bash	`#!/bin/bash`
Bourne shell	`#!/bin/sh`
Powershell	`#!/user/bin/pwsh`
Other scripting languages	`#!/user/bin/env <interpreter>`

Table 12.1 – Shebang lines for different interpreters

Adding comments

To add comments to your bash script, simply start the comment with the # symbol and the pound sign, followed by comment text. This text can be used by you and other developers to document information that might not be easily understood from the code alone. It could also just add some details on the usage of the script, who the author is, and so on.

Adding print lines

In a bash file, to print lines out, simply use the `echo` command. These print lines will help you to understand exactly where your application is within its running process. Use these lines with intention and they will give you and your users some useful insight that can even make debugging easier.

Adding code

Within the bash script, we'll generate builds for all the differing build tags for each operating system and architecture pair. Let's first start to see which architecture values are available for Darwin:

```
go tool dist list | grep darwin
```

The values returned are as follows:

```
darwin/amd64
darwin/arm64
```

Let's generate the different Darwin builds – free, pro, and profile versions – for all architectures with the following code:

```
# Generate darwin builds

darwin_archs=(amd64 arm64)
```

```
for darwin_arch in ${darwin_archs[@]}
do
    echo "building for darwin/${darwin_arch} free version..."
    env GOOS=darwin GOARCH=${darwin_arch} go build -tags free
-o builds/free/darwin/${darwin_arch}/audiofile main.go
    echo "building for darwin/${darwin_arch} pro version..."
    env GOOS=darwin GOARCH=${darwin_arch} go build -tags pro -o
builds/pro/darwin/${darwin_arch}/audiofile main.go
    echo "building for darwin/${darwin_arch} profile
version..."
    env GOOS=darwin GOARCH=${darwin_arch} go build -tags
profile -o builds/profile/darwin/${darwin_arch}/audiofile main.
go
done
```

Next, let's do the same with Linux, first grabbing the architecture values available:

```
go tool dist list | grep linux
```

The values returned are as follows:

```
linux/386          linux/mips64le
linux/amd64        linux/mipsle
linux/arm          linux/ppc64
linux/arm64        linux/ppc64le
linux/loong64      linux/riscv64
linux/mips          linux/s390x
linux/mips64
```

Let's generate the different Linux builds – the free, pro, and profile versions – for all architectures with the following code:

```
# Generate linux builds

linux_archs=(386 amd64 arm arm64 loong64 mips mips64 mips64le
mipsle ppc64 ppc64le riscv64 s390x)

for linux_arch in ${linux_archs[@]}
do
```

```
    echo "building for linux/${linux_arch} free version..."
    env GOOS=linux GOARCH=${linux_arch} go build -tags free -o
builds/free/linux/${linux_arch}/audiofile main.go
    echo "building for linux/${linux_arch} pro version..."
    env GOOS=linux GOARCH=${linux_arch} go build -tags pro -o
builds/pro/linux/${linux_arch}/audiofile main.go
    echo "building for linux/${linux_arch} profile version..."
    env GOOS=linux GOARCH=${linux_arch} go build -tags profile
-o builds/profile/linux/${linux_arch}/audiofile main.go
done
```

Next, let's do the same with Windows, first grabbing the architecture values available:

```
go tool dist list | grep windows
```

The values returned are as follows:

```
windows/386
windows/amd64
windows/arm
windows/arm64
```

Finally, let's generate the different Windows builds – the free, pro, and profile versions – for all architectures with the following code:

```
# Generate windows builds
windows_archs=(386 amd64 arm arm64)
for windows_arch in ${windows_archs[@]}
do
    echo "building for windows/${windows_arch} free version..."
    env GOOS=windows GOARCH=${windows_arch} go build -tags free
-o builds/free/windows/${windows_arch}/audiofile.exe main.go
    echo "building for windows/${windows_arch} pro version..."
    env GOOS=windows GOARCH=${windows_arch} go build -tags pro
-o builds/pro/windows/${windows_arch}/audiofile.exe main.go
    echo "building for windows/${windows_arch} profile
version..."
    env GOOS=windows GOARCH=${windows_arch} go build -tags
profile -o builds/profile/windows/${windows_arch}/audiofile.exe
main.go
done
```

Here's the code when run from the Darwin/macOS or Linux terminal:

```
./build.sh
```

We can check that the executable files have been generated. The full list is quite long, and they have been organized within the following nested folder structure:

```
/builds/{level}/{operating-system}/{architecture}/{audiofile-
executable}
```

Figure 12.2 – Screenshot of generated folders from the build bash script

A script to generate these builds will need to be different if run on Windows, for example. If you are running your application on Darwin or Linux, try running the build script and see the generated builds populate. You can now share these builds with other users running on a different platform. Next, we'll create a PowerShell script to generate the same builds to run in Windows.

Creating a PowerShell script in Windows

Let's start by creating a PowerShell script. Let's name it `build.ps1`. Create the file by typing the following command within PowerShell:

```
notepad build.ps1
```

The preceding command asks to create the file when it does not exist. The file extension is `.ps1`, which indicates that the file is a PowerShell script type. Next, we want to edit it. You may use Notepad or another editor of your choice.

Unlike a bash script, a PowerShell script does not require a shebang. To learn more about how to write a PowerShell script, you can review the documentation here: `https://learn.microsoft.com/en-us/powershell/`.

Adding comments

To add comments to your PowerShell script, simply start the comment with a # symbol and a pound sign, followed by comment text.

Adding print lines

In a PowerShell file, to print lines out, simply use the `Write-Output` command:

```
Write-Output "building for windows/amd64..."
```

Writing output will help you to understand exactly where your application is within its running process, make it easier to debug, and give the user a sense that something is running. Having no output at all is not only boring but also communicates nothing to the user.

Adding code

Within the PowerShell script, we'll generate builds for all the differing build tags for each operating system and architecture pair. Let's start by seeing which architecture values are available for Darwin via a Windows command:

```
PS C:\Users\mmontagnino\Code\src\github.com\marianina8\
audiofile> go tool dist list | Select-String darwin
```

Using the `Select-String` command, we can return only the values that contain `darwin`. These values are returned:

```
darwin/amd64
darwin/arm64
```

We can run a similar command for Linux:

```
PS C:\Users\mmontagnino\Code\src\github.com\marianina8\
audiofile> go tool dist list | Select-String linux
```

And a command for Windows:

```
PS C:\Users\mmontagnino\Code\src\github.com\marianina8\
audiofile> go tool dist list | Select-String windows
```

The same values are returned within the previous sections, so I won't print them out. However, now that we know how to get the architecture for each operating system, we can add the code to generate the builds for all of them.

The code to generate Darwin builds is as follows:

```
# Generate darwin builds
$darwin_archs="amd64","arm64"
foreach ($darwin_arch in $darwin_archs)
{
    Write-Output "building for darwin/$($darwin_arch) free
version..."
    $env:GOOS="darwin";$env:GOARCH=$darwin_arch; go build -tags
free -o .\builds\free\darwin\$darwin_arch\audiofile main.go
    Write-Output "building for darwin/$($darwin_arch) pro
version..."
    $env:GOOS="darwin";$env:GOARCH=$darwin_arch; go build -tags
pro -o .\builds\pro\darwin\$darwin_arch\audiofile main.go
    Write-Output "building for darwin/$($darwin_arch) profile
version..."
    $env:GOOS="darwin";$env:GOARCH=$darwin_arch; go build -tags
profile -o .\builds\profile\darwin\$darwin_arch\audiofile main.
go
}
```

The code to generate Linux builds is as follows:

```
# Generate linux builds
$linux_
archs="386","amd64","arm","arm64","loong64","mips","mips64",
"mips64le","mipsle","ppc64","ppc64le","riscv64","s390x"
foreach ($linux_arch in $linux_archs)
{
    Write-Output "building for linux/$($linux_arch) free
version..."
    $env:GOOS="linux";$env:GOARCH=$linux_arch; go build -tags
free -o .
\builds\free\linux\$linux_arch\audiofile main.go
    Write-Output "building for linux/$($linux_arch) pro
version..."
    $env:GOOS="linux";$env:GOARCH=$linux_arch; go build -tags
pro -o .
\builds\pro\linux\$linux_arch\audiofile main.go
    Write-Output "building for linux/$($linux_arch) profile
version..."
    $env:GOOS="linux";$env:GOARCH=$linux_arch; go build -tags
profile -o
.\builds\profile\linux\$linux_arch\audiofile main.go
}
```

Finally, the code to generate Windows builds is as follows:

```
# Generate windows builds
$windows_archs="386","amd64","arm","arm64"
foreach ($windows_arch in $windows_archs)
{
    Write-Output "building for windows/$($windows_arch) free
version..."
    $env:GOOS="windows";$env:GOARCH=$windows_arch; go build
-tags free -o .\builds\free\windows\$windows_arch\audiofile.exe
main.go
    Write-Output "building for windows/$($windows_arch) pro
version..."
    $env:GOOS="windows";$env:GOARCH=$windows_arch; go build
-tags pro -o .\builds\pro\windows\$windows_arch\audiofile.exe
main.go
```

```
    Write-Output "building for windows/$($windows_arch) profile
version..."

    $env:GOOS="windows";$env:GOARCH=$windows_arch; go build
-tags profile -o .\builds\profile\windows\$windows_arch\
audiofile.exe main.go
}
```

Each section generates a build for one of the three major operating systems and all the available architectures. To run the script from PowerShell, just run the following script:

./build.ps1

The following will be the output for each port:

building for $GOOS/$GOARCH [free/pro/profile] version...

Check the builds folder to see all the ports generated successfully. The full list is quite long, and they have been organized within the following nested folder structure:

**/builds/{level}/{operating-system}/{architecture}/{audiofile-
executable}**

Now, we can generate builds for all operating systems and architectures from a PowerShell script, which can be run on Windows. If you run any of the major operating systems – Darwin, Linux, or Windows – you can now generate a build for your own platform or anyone else who would like to use your application.

Summary

In this chapter, you learned what the GOOS and GOARCH environment variables are and how you can use them, as well as build tags, to customize builds based on the operating system, architecture, and levels. These environment variables help you to learn more about the environment you're building in and possibly understand why a build may have trouble executing on another platform.

There are also two ways to compile an application – building or installing. In this chapter, we discussed how to build or install the application and what the difference is. The same flags are available for each command, but we discussed how to build or install on each of the major operating systems using the Makefile. However, this also showed how large the Makefile can become!

Finally, we learned how to create a simple script to run in Darwin, Linux, or Windows to generate all the builds needed for all the major operating systems. You learned how to write both a bash and PowerShell script to generate builds. In the next chapter, *Chapter 13*, *Using Containers for Distribution*, we will learn how to run these compiled applications on containers made from different operating system images. Finally, in *Chapter 14*, *Publishing Your Go Binary as a Homebrew Formula with*

GoReleaser, you'll explore the tools required to automate the process of building and releasing your Go binaries across a range of operating systems and architectures. By learning how to use GoReleaser, you can significantly accelerate the process of releasing and deploying your application. This way, you can concentrate on developing new features and addressing bugs instead of getting bogged down with the build and compile process. Ultimately, using GoReleaser can save you valuable time and energy that you can use to make your application even better.

Questions

1. What Go environment variables define the operating system and the architecture?
2. What additional security do you get from building with a first-class port?
3. What command would you run on Linux to find the port values for the Darwin operating system?

Answers

1. GOOS is the Golang operating system, and GOARCH is the Golang architecture value.
2. There are several reasons why a first-class port is more secure: releases are blocked by broken builds, official binaries are provided, and installation is documented.
3. `go tool dist list | grep darwin`.

Further reading

*   Read more about compiling at `https://go.dev/doc/tutorial/compile-install`
*   Read more about Go environment variables at `https://pkg.go.dev/cmd/go`

13

Using Containers for Distribution

In this chapter, we'll explore the world of containerization and examine the many reasons why you should use Docker containers for testing and distributing your applications. The term *containerization* refers to a style of software packaging that makes it simple to deploy and run in any setting. First, we'll go over the basics of Docker, covered by a simple application that can be built into an image and run as a container. Then, we return to our audiofile application, for a more advanced example, to learn how to create multiple Docker containers that can be composed and run together. These examples give you not only an understanding of the basic flags used for running containers but also some advanced flags that show you how to run containers with mapped network stacks, volumes, and ports.

We also explain how to use Docker containers for integration testing, which increases your confidence, because, let's face it, mocking API responses can cover only so much. A good mix of unit and integration tests gives you not just the coverage but also confidence that the overall system works.

Finally, we will discuss some of the disadvantages of adopting Docker. Consider the increased complexity of administering containerized applications, as well as the additional overhead of operating several containers on a single host. Docker as an external dependency may be a disadvantage in itself. This chapter will help you determine when to use, and not to use, containers for your application.

By the end of this chapter, you will have a strong grasp of how to utilize Docker containers and how they might assist your development, testing, and deployment workflow. You will be able to containerize your application, test it with Docker, and release it with Docker Hub. Specifically, we'll cover the following topics:

- Why use containers?
- Testing with containers
- Distributing with containers

Technical requirement

For this chapter, you will need to do the following:

- Download and install Docker Desktop at `https://www.docker.com/products/docker-desktop/`
- Install the Docker Compose plugin

You can also find the code examples on GitHub at `https://github.com/PacktPublishing/Building-Modern-CLI-Applications-in-Go/tree/main/Chapter13`

Why use containers?

First, let's talk about what a container is. A **container** is a standardized software unit that allows the transport of a program from one computing environment quickly and reliably to another by bundling the application's code and all its dependencies into a single encapsulation. Simply put, containers let you package all your dependencies into a single container so that it can run on any machine. Containers are isolated from one another and bundle their own system libraries and settings, so they don't conflict with other containers or the host system. This makes them a lightweight and portable alternative to **virtual machines** (VMs). Popular containerization tools include **Docker** and **Kubernetes**.

Benefiting from containers

Let's break down some of the benefits of using containers in your Go project:

- **Portability**: Containers make it possible to support consistency of behavior across various environments, lowering the possibility of errors and incompatibilities.
- **Isolation**: They offer a degree of isolation from the host system and other containers, which increases their level of security and reduces their propensity for conflicts.
- **Lightweight**: Compared to VMs, containers are smaller and start up faster, which increases their operating efficiency.
- **Scalability**: They can be easily scaled up or down, enabling effective resource use. For example, if you utilize containers for your application, then you can create multiple identical containers running your application deployed across multiple servers.
- **Versioning**: Containers can be versioned, making it simple to revert to earlier iterations as needed.
- **Modularity**: Because containers can be created and managed separately, they are simple to update and maintain.
- **Cost-effective**: By lowering the number of systems you need to run your applications, containers can help you save money on infrastructure and maintenance.

Creating and running command-line applications is made simple and reliable by containers. Regardless of the host machine's configuration, this means that the application will always be built and run in the same manner. Application development and deployment across different operating systems are made significantly simpler by including all necessary dependencies and runtime environments within the container image. Finally, containers make it simple to duplicate development environments, enabling multiple developers or teams to work together in the same area while guaranteeing that the application is developed and executed uniformly across various environments.

Additionally, using containers makes it simpler to integrate applications with **continuous integration and continuous deployment (CI/CD)** pipelines. Since all the necessary dependencies exist within the container's image, the pipeline can more reliably and easily build and run the application, eliminating the need to configure the pipeline's host machine's development environment.

Finally, the consistency of an isolated environment with containers is another benefit that makes it easier to distribute your command-line application while guaranteeing that it will operate exactly as expected. Users no longer need to configure their environment for the application, making containers, while also lightweight, a great way to distribute across various environments and platforms.

As you can clearly see, there are a variety of situations where containers can prove useful, including command-line application development and testing! Now, let's discuss when you may not want to use containers.

Deciding not to use containers

While containers are often helpful, there are some circumstances in which they might not be the best option:

- **High-performance computing**: High-performance computing and other tasks that need direct access to the host system's resources might not be good candidates for containers because of the additional overhead they cause.

- **Requiring high levels of security**: Containers share the host's kernel and might not offer as much isolation as a VM. VMs may be a better option if your workload demands a high level of security.

- **Neglecting container-native features**: You may not see the benefit of using containers if you do not plan on using any of the native features included for scaling, rolling updates, service discovery, and load balancing.

- **Inflexible applications**: If an application requires a very specific operating environment in order to function properly, it might not even be possible to containerize it, as there are limited operating systems and platforms that are supported.

- **Team inertia**: If you or your team are unwilling to learn about containers and container orchestration, then it will be difficult to incorporate a new tool.

Nevertheless, it's important to note that these situations are not always the case and that there are some solutions available, including the use of VMs, particular security features of container orchestration platforms, specialized container runtimes such as **gVisor** or **Firecracker**, and others.

In the following examples and within the next sections, we will be using Docker to show how easy it can be to start using Docker and use it to create a consistent environment for testing and distribution.

In the Chapter-13 GitHub repository, we go over a very simple example for building an image and running a container. The main.go file is simple:

```go
func main() {
    var helloFlag bool
    flag.BoolVar(&helloFlag, "hello", false, "Print 'Hello,
      World!'")
    flag.Parse()
    if helloFlag {
        fmt.Println("Hello, World!")
    }
}
```

Passing in the hello flag to the built application will print out "Hello, World!".

Building a simple Docker image

To start, software can be packaged as an **image**, a small, self-contained executable that contains the program's source code, libraries, configuration files, runtime, and environment variables. Images are the building blocks of containers and are used to create and run them.

Let's build a Docker image for this very simple application. To do so, we'll need to create a **Dockerfile**. You can create a file named Dockerfile that will automatically be recognized when you run the command-line Docker commands, or create a file with the .dockerfile extension, which will require the -f or --file flag for passing in the filename.

A Dockerfile contains instructions for building a Docker image, as depicted in the following diagram. Each instruction creates a new layer within the image. The layers are combined to create the final image. There are many different kinds of instructions you can put in the Dockerfile. For example, you can tell Docker to copy files into the base image, set environment variables, run commands, and specify the executables to run when a container is initialized:

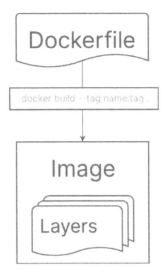

Figure 13.1 – Visual of a Dockerfile transformed into an image with layers by the build command

For our base image, let's visit Docker Hub's website at https://hub.docker.com and search for the official Golang Docker base image for Go v1.19. We see that we can use the golang image with tag 1.19. The FROM instruction is the first line of the Dockerfile and it sets the base image to use:

```
FROM golang:1.19
```

Then, copy all the files:

```
COPY . .
```

Build the hello world application:

```
RUN go build main.go
```

Finally, run the application while passing the hello flag:

```
CMD ["./main", "--hello"]
```

Altogether, the Dockerfile contains the preceding instructions with some descriptive comments indicated by # as the first character in line.

To build a Docker image from a Dockerfile, we call the docker build command. The command takes the following syntax:

```
docker build [options] path| url | -
```

When run, the command does the following:

- Reads the instructions specified within the Dockerfile and performs them in order

- Each instruction creates a new layer in the image, and the final image combines them all

- Tags the new image with the specified or generated name, and—optionally—a tag in the name:tag format

The options parameter can be used to pass in different options to the command, which can include build-time variables, targets, and more. The path | url | - argument specifies the location of the Dockerfile.

Let's try building this image from the Dockerfile we created for our hello world application. Within the root of the repository, run the following command:

```
docker build --tag hello-world:latest
```

After running the command, you should see similar output to this:

```
[+] Building 2.4s (8/8) FINISHED
=> [internal] load build definition from
Dockerfile            0.0s
=> => transferring dockerfile:
238B                         0.0s
=> [internal] load
.dockerignore                            0.0s
=> => transferring context:
2B                        0.0s
=> [internal] load metadata for docker.io/library/golang:1.19
1.2s
=> [internal] load build
context                   0.0s
=> => transferring context:
2.25kB                      0.0s
=> CACHED [1/4] FROM docker.io/library/golang:1.19@
sha256:bb9811fad43a7d6fd217324 0.0s
=> [2/4] COPY .
.                               0.0s
=> [3/4] RUN go build main.
go                       1.0s
=> exporting to
image                        0.1s
=> => exporting
```

```
layers                                              0.0s
=> => writing image
sha256:91f97dc0109218173ccae884981f700c83848aaf524266de20f950
  0.0s
=> => naming to docker.io/library/hello-world:latest
0.0s
```

From about midway through the output, you'll see that the layers of the image are built, concluding with the final image tagged as `hello-world:latest`.

You can view the images that exist, by running the following command in your terminal:

```
% docker images
REPOSITORY          TAG        IMAGE ID        CREATED
SIZE
hello-world       latest     91f97dc01092    18 minutes ago   846MB
```

Now that we've successfully built our Docker image for this simple hello world application, let's follow up by running it within a container.

Running a simple Docker container

When you run a Docker container, Docker Engine takes an existing image and creates a new running instance of it. This container exists within an isolated environment that has its own filesystem, network interfaces, and process space. However, the image is a necessary starting point for creating—or running—the container.

> **Note**
>
> When a container is running, it can make changes to the filesystem, such as creating or modifying files. However, these changes are not saved in the image and will be lost when the container is stopped. If you want to save the changes, you can create a new image of the container using the `docker commit` command.

To create and run a Docker container from an image, we call the `docker run` command. The command takes the following syntax:

```
docker run [options] image[:tag] [command] [arg...]
```

The `docker run` command checks if the image exists locally; if not, then it will pull it from Docker Hub. Docker Engine then creates a new container from this image, with all layers or instructions applied. We'll break this down here:

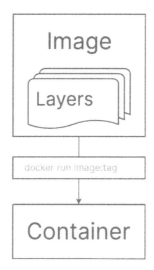

Figure 13.2 – Visual of an image used to create a container with the run command

As mentioned, when `docker run` is called, the following steps occur:

1. Docker checks if the requested image exists locally; if not, it retrieves it from a registry, such as Docker Hub.

2. From the image, it creates a new container.

3. It starts the container and executes the commands specified within the instructions of the Dockerfile.

4. It attaches the terminal to the container's process in order to display any output from the commands.

The `options` parameter can be used to pass in different options to the command, which can include mapping ports, setting environment variables, and more. The `image[:tag]` argument specifies the image to use for creating the container. Finally, the `command` and `[arg...]` arguments are used to specify any commands to run within the container.

In each example that we call the `docker run` command, we pass in the `--rm` flag, which tells Docker to automatically remove the container when it exits. This will save you from accidentally ending up with many gigabytes of stopped containers sitting in the background.

Try to run an image from the `hello-world:latest` image that we created for our hello world application. Within the root of the repository, run the following command and see the text output:

```
% docker run --rm hello-world:latest
Hello, World!
```

We did it! A simple Dockerfile for a simple hello world application. Within the next two sections, we'll return to the audiofile command-line application example and use this new skill of building images and running containers for testing and distribution.

Testing with containers

So far, within our command-line application journey, we've built tests and mocked the service output. The benefit of using containers, besides having a consistent and isolated environment for tests to run on any host machine, is that you can use them for running integration tests that provide more reliable test coverage for your application.

Creating the integration test file

We created a new `integration_test.go` file to handle the configuration and execution of integration tests, but we don't want it to run with all the other tests. To specify its uniqueness, let's tag it with `int`, short for integration. At the top of the file, we add the following build tag:

```
//go:build int && pro
```

We include the `pro` build tag because we are testing all the available features.

Writing the integration tests

First, let's write the `ConfigureTest()` function to prepare for our integration tests:

```
func ConfigureTest() {
    getClient = &http.Client{
        Timeout: 15 * time.Second,
    }
    viper.SetDefault("cli.hostname", "localhost")
    viper.SetDefault("cli.port", 8000)
    utils.InitCLILogger()
}
```

In the preceding code, you can see that we use an actual client, not the mocked client that is currently used within unit tests. We use `viper` to set the hostname and port for the API we connect as localhost on port `8000`. Finally, we initialize the logger files so that we don't run into any panics while logging.

For the integration test, let's use a specific workflow:

1. **Upload audio**: First, we want to make sure an audio file exists within the local storage.

2. **Get audio by id**: From the previous step, we can retrieve the audiofile ID returned and use this to retrieve the audio metadata from storage.

3. **List all audio**: We list all the audio metadata and confirm that the previously uploaded audio exists within the list.

4. **Search audio by value**: Search for that uploaded audio based on metadata we know exists within the description.

5. **Delete audio by id**: Finally, delete the initial audio file we uploaded by the ID we retrieved from *step 1*.

The order is specific as the latter steps within the workflow depend on the first.

The integration tests are like the unit tests, but paths to real files are passed in, and the actual API is called. Within the `integration_tests.go` file exists a `TestWorkflow` function that calls the commands in the order listed previously. Since the code is similar to the unit tests, let's just go over the first two command calls, and then move straight into using Docker to execute the integration tests!

Before any methods are tested, the integration test is configured by calling the `ConfigureTest` function:

```
ConfigureTest()
fmt.Println("*** Testing upload ***")
b := bytes.NewBufferString("")
rootCmd.SetOut(b)
rootCmd.SetArgs([]string{"upload", "--filename",
   "../audio/algorithms.mp3"})
err := rootCmd.Execute()
if err != nil {
    fmt.Println("err: ", err)
}
uploadResponse, err := ioutil.ReadAll(b)
if err != nil {
    t.Fatal(err)
}
id := string(uploadResponse)
if id == "" {
    t.Fatalf("expected id returned")
}
```

In the preceding code, we then use rootCmd to call the upload command with the filename set to ../audio/algorithms.mp3. We execute the command and read the response back as a byte slice that is then converted to a string and stored in the id variable. This id variable is then used for the following tests. We run the get command and pass in the same id variable to retrieve the audiofile metadata for the previously uploaded audio:

```
fmt.Println("*** Testing get ***")
rootCmd.SetArgs([]string{"get", "--id", id, "--json"})
err = rootCmd.Execute()
if err != nil {
    fmt.Println("err: ", err)
}
getResponse, err := ioutil.ReadAll(b)
if err != nil {
    t.Fatal(err)
}
var audio models.Audio
json.Unmarshal(getResponse, &audio)
if audio.Id != id {
    t.Fatalf("expected matching audiofile returned")
}
```

We continue testing the list, search, and delete commands similarly and ensure that the specific metadata with a matching id variable is returned each time. When the tests are done, we try to run the integration test. Without the API running locally, running the following command fails miserably:

```
go test ./cmd -tags "int pro"
```

Before we try again, let's build a Dockerfile to run the API within a contained environment.

Writing the Dockerfiles

In the real world, our API might be hosted on some external website. However, we are currently running on localhost, and running it within a container will allow users to easily run it no matter which machine they use. In this section, we will create two Dockerfiles: one for the CLI and another for the API.

Writing the API Dockerfile

First, we'll create an api.Dockerfile file to hold all the instructions to build the image and run the container for the audiofile API:

```
FROM golang:1.19

# Set the working directory
WORKDIR /audiofile

# Copy the source code
COPY . .

# Download the dependencies
RUN go mod download

# Expose port 8000
EXPOSE 8000

# Build the audiofile application with the pro tag so all
# features are available
RUN go build -tags "pro" -o audiofile main.go
RUN chmod +x audiofile

# Start the audiofile API
CMD ["./audiofile", "api"]
```

Let's build this image. The -f flag allows you to specify the api.Dockerfile file to use, and the -t flag allows you to name and tag the image:

```
% docker build -f api.Dockerfile -t audiofile:api .
```

After the command executes, you can run the docker images command to confirm its creation:

```
% docker images
REPOSITORY        TAG        IMAGE
ID        CREATED        SIZE
audiofile        api        12afba7f3fb7        9 minutes
ago   1.75GB
```

Now that we see that the image has been built successfully, let's run the container and test it out!

Run the following command to run the container:

```
% docker run -p 8000:8000 --rm audiofile:api
Starting API at http://localhost:8000
Press Ctrl-C to stop.
```

You'll see the preceding output if the API is started successfully. We have the audiofile API running in a container within your host. Remember that any commands will check against the flat file storage, pointing to the `audiofile` directory created under the `home` directory. Any audio files uploaded, processed, and with metadata stored within the container will not be saved unless we commit the changes. Since we are just running integration tests, this won't be necessary.

> **Note**
> The `-p` flag within the `docker run` command allows you to specify the port mapping between the host and container. The syntax is `-p host_port:container_port`. This maps the host's port to the container's port.

Within a separate terminal, let's run the integration tests again and see them pass:

```
% go test ./cmd -tags "int pro"
ok      github.com/marianina8/audiofile/cmd    0.909s
```

Success! We've now run integration tests connecting to the audiofile API within a container.

Writing the CLI Dockerfile

Now, for running the CLI integration tests within a container, we'll create a `cli.Dockerfile` file. It will hold all the instructions to build the image and run the container for the integration tests:

```
FROM golang:1.19

# Set the working directory
WORKDIR /audiofile

# Copy the source code
COPY . .

# Download the dependencies
RUN go mod download
```

```
# Execute `go test -v ./cmd -tags int pro` when the
# container is running
CMD ["go", "test", "-v", "./cmd", "-tags", "int pro"]
```

The preceding comments clarify each instruction, but let's break down the Docker instructions:

1.   Specify and pull from the base image as `golang:1.19`.

2.   Set the working directory to `/audiofile`.

3.   Copy over all the source code to the working directory.

4.   Download all the Go dependencies.

5.   Execute `go test -v ./cmd -tags int pro`.

Let's build this image:

% docker build -f cli.Dockerfile -t audiofile:cli .

Then, while ensuring the `audiofile:api` container is already running, run the `audiofile:cli` container:

% docker run --rm --network host audiofile:cli

You'll see that the integration tests run successfully.

> **Note**
>
> The `--network host` flag within the `docker run` command is used to connect a container to the host's network stack. It means that the container will have access to the host's network interfaces, IP address, and ports. Be careful with security if the container runs any service.

Now, we've created two containers for the API and CLI, but rather than having to run each separately within two separate terminals, it'd be easier to use **Docker Compose**. Docker Compose is a plugin for Docker Engine that allows you to define and run multiple Docker applications all with a single file, `docker-compose.yml`, starting and stopping the entire application with a single `stop/ start` command.

Writing the Docker Compose file

Inside the `docker-compose.yml` Docker Compose file, we define both containers that need to be run, while specifying any parameters we've previously set via flags for the `docker run` command:

```
version: '3'
services:
```

```
cli:
  build:
    context: .
    dockerfile: cli.Dockerfile
  image: audiofile:cli
  network_mode: host
  depends_on:
    - api
api:
  build:
    context: .
    dockerfile: api.Dockerfile
  image: audiofile:api
  ports:
  - "8000:8000"
```

Let's explain the preceding file. First, there are two services defined: `cli` and `api`. Beneath each service are a set of similar keys:

- The `build` key, which is used to specify the context and location of the Dockerfile.

- The `context` key is used to specify where to look for the Dockerfile. Both are set to `.`, which tells the Docker Compose service to look in the current directory.

- The `dockerfile` key allows us to specify the name of the Dockerfile—in this case, `cli.Dockerfile` for the `cli` service and `api.Dockerfile` for the `api` service.

- The `image` key allows us to give a name and tag the image.

For the `cli` service, we've added some further keys:

- The `network_mode` key is used to specify the networking mode for a service. When it is set to `host`, like it is for the `cli` service, it means to use the host machine's network stack (like the `–network host` flag used when calling `docker run` for the CLI).

- The `depends_on` key allows us to specify the order of which services should be running first. In this case, the `api` service must be running first

- For the `api` service, there's an additional key:

- The `ports` key is used to specify port mappings between the host machine and the container. Its syntax is `` `host_port:container_port` `` and is like the `-p` or `--publish` flag when calling the `docker run` command.

Now that we've got the Docker Compose file completed, we just have one simple command, `docker-compose up`, to run the integration tests within a containerized environment:

```
% docker-compose up
[+] Running 3/2
  Network audiofile_
  default  Created                        0.1s
  Container audiofile-
  api-1  Created                          0.0s
  Container audiofile-
  cli-1  Created                          0.0s
Attaching to audiofile-api-1, audiofile-cli-1
audiofile-api-1  | Starting API at http://localhost:8000
audiofile-api-1  | Press Ctrl-C to stop.
audiofile-cli-1  | === RUN   TestWorkflow
audiofile-cli-1  | --- PASS: TestWorkflow (1.14s)
...
audiofile-cli-1  | ok   github.com/marianina8/audiofile/
cmd      1.163s
```

Now, no matter which platform you're running the containers on, the results will be consistent while running the tests within a container. Integration testing provides more comprehensive testing as it will catch bugs that might exist within the **end-to-end** (**E2E**) flow from the command to the API to the filesystem and back. We can therefore increase our confidence with tests that can ensure our CLI and API are more stable and reliable as a whole. In the next section, we'll discuss how to distribute your CLI application with containers.

Distributing with containers

There are various advantages to running a CLI inside a container as opposed to directly on the host. Utilizing a container makes the setup and installation of the program easier. This can be helpful if the application needs numerous dependencies or libraries that are challenging to install. Additionally, regardless of the language or tools used to construct the program, adopting a container enables a more dependable and uniform method of distribution. Using a container as a distribution method can be a flexible solution for the majority of applications that can operate in a Linux environment, even though there may be language-specific alternatives. Finally, distributing through containers will be useful for developers unfamiliar with the Go language but who already have the Docker toolbox installed on their machines.

Building a new image to run as an executable

To build an image that can run as an executable, we must create an ENTRYPOINT instruction on the image to specify the main executable. Let's create a new Dockerfile, dist.Dockerfile, which contains the following instructions:

```
FROM golang:1.19

# Set the working directory
WORKDIR /audiofile

# Copy the source code
COPY . .

# Download the dependencies
RUN go mod download

# Expose port 8000
EXPOSE 8000

# Build the audiofile application with the pro tag so all
# features are available
RUN go build -tags "pro" -o audiofile main.go

# Start the audiofile API
ENTRYPOINT ["./audiofile"]
```

Since these instructions are mostly similar to the other Dockerfiles explained in the previous sections, we won't go into a detailed explanation. The only instruction to note is the ENTRYPOINT instruction, which specifies ./audiofile as the main executable.

We can build this image with the following command:

```
% docker build -f dist.Dockerfile -t audiofile:dist .
```

After confirming that the image is successfully built, we are now ready to run the container and interact with it as an executable.

Interacting with your container as an executable

To interact with your container as an executable, you can configure your container to use an interactive TTY (terminal) with the ENTRYPOINT command in Docker. The -i and -t options stand for *interactive* and *TTY* respectively, and when the two flags work together, you can interact with the ENTRYPOINT command in a terminal-like environment. Remember to have the API running first. Now, let's show how it'll look when we run the container for the audiofile:dist image:

```
% docker run --rm --network host -ti audiofile:dist help
A command line interface allows you to interact with the
Audiofile service.
Basic commands include: get, list, and upload.

Usage:
  audiofile [command]

Available Commands:
  ...


Use "audiofile [command] --help" for more information about a
command.
```

Just by typing help at the end of the docker run command passes in help as input to the main executable, or ENTRYPOINT: ./audiofile. As expected, the help text is output.

The docker run command uses a few additional commands; the -network host flag uses the host's network stack for the container, and the -rm command tells Docker to automatically remove the container when it exits.

You can run any of the commands by just replacing the word help with the name of the other command. To run upload, for example, run this command:

```
% docker run --rm --network host -ti audiofile:dist upload -
filename audio/algorithms.mp3
```

You can now interact with your command-line application through a container passing in commands and not have to worry if it will respond any differently based on the host machine. As previously mentioned, any filesystem changes, or files uploaded, as in preceding the file, are not saved when the container exists. There is a way to run the API so that local file storage maps to a container path.

Mapping host machine to container file paths

As mentioned, you can map a host machine path to a Docker container file path so that files on the host computer may be accessed from inside the container. This can be helpful for things such as giving the container access to data volumes or application configuration files.

The -v or –volume option can be used to translate a host machine path to a container path when executing a container. This flag's syntax is host path:container path. For instance, the docker run -v /app/config:/etc/config imageName:tag command would be used to map the host machine's /app/config directory to the container's /etc/config directory.

It's crucial to remember that both the host path and container path need to be present in the container image before the container can be executed. You must construct the container path before starting the container if it does not already exist in the container image.

If you dig into the audiofile API that is running on your local host, you'll see that the flat file storage is mapped to the /audiofile folder existing under the host's home directory. On my macOS instance, if I wanted to run the audiofile API within a Docker container but be able to read from and access or upload data to the flat file storage, then I would need to map the audiofile directory under my HOME directory to an appropriate location. This docker run command would do it:

```
docker run -p 8000:8000 --rm -v $HOME/audiofile:/root/
audiofile  audiofile:api
```

Run the preceding command first and then run the CLI container, or modify the docker-compose. yml file's API service to include the following:

```
volumes:
  - "${HOME}/audiofile:/root/audiofile"
```

Either way, when you run the container for integration tests or as an executable, you'll be interacting with your local storage mapped to the /root/audiofile directory within the container. If you've been playing around with the audiofile CLI and uploading directory, then when you start the container up and run the list command, you'll see preexisting metadata instead of an empty list returned.

Mapping a path from your host to a container is an option that you can share with your users when instructing them how to use the audiofile application.

Reducing image size by using multi-stage builds

By running the docker images command, you'll see that some of the images built are quite large. To reduce the size of these images, you may need to rewrite your Dockerfile to use multi-stage builds. A **multi-stage build** is a process of dividing up the build into multiple stages, in which it is possible to remove unnecessary dependencies, artifacts, and configurations from the final image. This is especially useful when building images for large applications where you can save on deployment time as well as infrastructure costs.

A way that single-stage and multi-stage builds differ is that multi-stage builds allow you to use multiple FROM statements, each defining a new stage of the build process. You can selectively copy artifacts, or builds, from one stage or another, allowing you to take what you need and discard the rest, essentially allowing you to remove anything unnecessary and clean up space.

Let's consider the `dist.Dockerfile` file and rewrite it. In our multi-stage build process, let's define our stages:

- **Stage 1**: Build our application
- **Stage 2**: Copy the executable, expose the port, and create an entry point

First, we create a new file, `dist-multistage.Dockerfile`, with the following instructions:

```
# Stage 1
FROM golang:1.19 AS build
WORKDIR /audiofile
COPY . .
RUN go mod download
RUN go build -tags "pro" -o audiofile main.go

# Stage 2
FROM alpine:latest
COPY --from=build /audiofile/audiofile .
EXPOSE 8000
ENTRYPOINT ["./audiofile"]
```

In *Stage 1*, we copy all the code files, download all dependencies, then build the application—basically, all as in the original instructions within `dist.Dockerfile`, but without the EXPOSE and ENTRYPOINT instructions. One thing to note is that we've named the stage `build`, with the following line:

```
FROM golang:1.19 AS build
```

In *Stage 2*, we copy over just the compiled binary from the `build` stage and nothing else. To do this, we run the following instruction:

```
COPY --from=build /audiofile/audiofile .
```

The command allows us to copy a file or directory from a previous stage, `build`, to the current stage. The `--from=build` option specifies the stage name to copy the file from. `/audiofile/audiofile` is the path of the file in the `build` stage, and `.` at the end of the command specifies the destination directory, the root directory, of the current stage.

Let's try building it and comparing the new size against the original size:

```
REPOSITORY  TAG               IMAGE ID        CREATED        SIZE
audiofile    dist                1361cbc7be3e    2 minutes
ago    1.78GB
audiofile      dist-multistage     ab5640f99ef2     5 minutes
ago    24MB
```

That's a big difference! Using multi-stage builds will help you to save on deployment time and infrastructure costs, so it's definitely worth the time writing your Dockerfiles using this process.

Distributing your Docker image

There are many methods for making your Docker images accessible to others. **Docker Hub**, a public registry where you can post your images and make them readily available to others, is a popular alternative. Another alternative is to use **GitHub Packages** to store and distribute your Docker images alongside other sorts of packages. There are other cloud-based registries such as **Amazon Elastic Container Registry** (ECR), **Google Container Registry** (GCR), and **Azure Container Registry** (ACR) that provide extra services such as image scanning (for OS vulnerabilities, for example) and signing.

It's a good idea to give instructions on how to utilize your image and run a container in the README file of the repository where the image is located. People who are interested in utilizing your image will be able to readily access instructions on how to retrieve the image, run a container using the image, and any other pertinent facts.

There are several advantages to publishing a Docker image, including simple distribution, versioning, deployment, collaboration, and scalability. Your image may be rapidly and readily distributed to others, making it simple for others to utilize and operate your application. Versioning helps you to maintain track of several versions of your image so that you may revert to an earlier version if necessary. Easy deployment allows you to deploy your application to several environments with little modifications. Sharing images via a registry facilitates collaboration with other developers on a project. And scalability is simple to accomplish by using the same image to create as many containers as you need, making it simple to grow your application.

In this chapter, we'll publish to Docker Hub as an example for our audiofile CLI project.

Publishing your Docker image

To publish an image to Docker Hub, you'll first need to create an account on the website. Once you have an account, you can sign in and create a new repository to store your image. After that, you can use the Docker command-line tool to log in to your Docker Hub account, tag your image with the repository name, and push the image to the repository. Here is an example of the commands you would use to do this:

```
docker login --username=your_username
docker tag your_image your_username/your_repository:your_tag
docker push your_username/your_repository:your_tag
```

1. Let's try this with our audiofile API and CLI images. First, I will log in with my username and password:

    ```
    % docker login --username=marianmontagnino
    Password:
    Login Succeeded

    Logging in with your password grants your terminal
    complete access to your account.
    For better security, log in with a limited-privilege
    personal access token. Learn more at https://docs.docker.
    com/go/access-tokens/
    ```

2. Next, I tag my CLI image:

    ```
     % docker tag audiofile:dist  marianmontagnino/
    audiofile:latest
    ```

3. Finally, I publish the image to Docker Hub:

    ```
    % docker push marianmontagnino/audiofile:latest
    The push refers to repository [docker.io/
    marianmontagnino/audiofile]
    c0f557e70e4f: Pushed
    98f8be277d74: Pushed
    6c199763ccbe: Pushed
    8f2f7ffa843f: Pushed
    10bb928a2e24: Pushed
    f1ce3f3654c3: Mounted from library/golang
    3685241d2bbb: Mounted from library/golang
    ```

```
dddbac67c6fa: Mounted from library/golang
85f9ebffaf4d: Mounted from library/golang
72235aad06ad: Mounted from library/golang
5d37ad02a8e2: Mounted from library/golang
ea8ab45f064e: Mounted from library/golang
latest: digest:
sha256:b7b3f58da01d360fc1a3f2e2bd617a44d3f7be
d6b6625464c9d787b8a71ead2e size: 2851
```

Let's confirm in Docker Hub to make sure that the container exists:

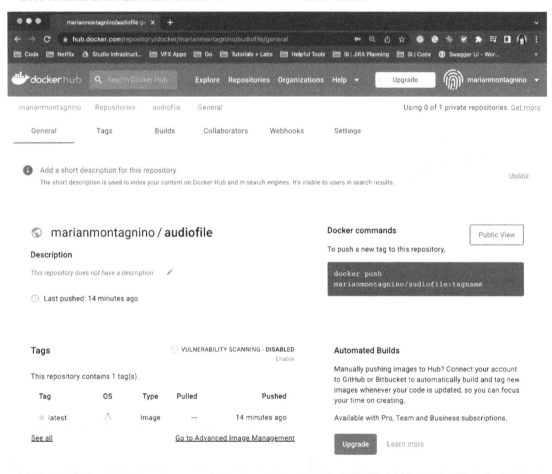

Figure 13.3 – Screenshot of the Docker Hub website showing the audiofile image tagged with latest

It's a great idea to include instructions for running a container using the image within a README file of the repository where the image is stored. This makes it easy for people who want to use the image to learn how to pull the image and run the container properly. As an example, here are sample instructions for our previously uploaded audiofile CLI image:

To run the audiofile CLI container, ensure that the audiofile API container is running first. Next, run the docker command:

```
% docker run --rm --network host -ti marianmontagnino/
audiofile:latest help
```

You'll see that the help text is output. Let's update the instructions on the Docker Hub repository.

Updating the README

From the Docker Hub repository, where our image is stored (in this example, the audiofile repository), we can scroll down to the bottom of the page to see a **README** section:

Figure 13.4 – Screenshot of the README section in the repository on Docker Hub

Click **here** to edit the repository description. Add the instructions we discussed previously, then click the **Update** button:

README

To run the audiofile CLI container, ensure that the audiofile API container is running first. Next, run the docker command:

```
docker run --rm --network host -ti marianmontagnino/audiofile:latest help
```

Figure 13.5 – Screenshot of the updated README section

Follow these instructions to similarly publish the audiofile API image to your Docker Hub repository. Now that the images exist in a public repository on Docker Hub, they are available to share and distribute to other users.

Depending on Docker

The fact that users must have Docker installed on their computers is one of the key disadvantages of utilizing Docker to deploy a CLI. However, if your program has complicated dependencies or is designed to operate on various platforms, this Docker dependency may be easier to handle. Using Docker may assist in avoiding difficulties with many libraries and unexpected interactions with various system setups.

Summary

We've gone into the realm of containerization and examined the numerous advantages of utilizing Docker containers for your applications in this chapter. The fundamentals of creating and running a simple Docker image and container are explained, as well as some more sophisticated instances using our audiofile application, which requires the construction of multiple containers that can be composed and run together.

Clearly, utilizing Docker for integration testing boosts your trust in the whole system, and we discussed how to run integration tests using Docker Compose.

At the same time, we've acknowledged some of Docker's drawbacks, such as the increased complexity of maintaining containerized applications, the additional burden of operating several containers on a single host, and the external dependency of Docker itself.

Overall, this chapter has given you a strong knowledge of when to utilize Docker containers for command-line applications—for testing and distribution. Now, you can ensure that the application runs consistently across any host machine. It is up to you, though, to decide if the upsides outweigh the downsides of having an external dependency and some level of complexity.

In the next chapter, *Chapter 14, Publishing your Go Binary as a Homebrew Formula with GoReleaser*, we'll take distribution to a next level. We will get your application available on the official Homebrew repository to further increase the distribution of your application.

Questions

1. Which command is used to create and run a container from an image?
2. Which `docker run` flag is used to attach a host machine path to a container path?
3. Which Docker command is used to see all created containers?

Further reading

- *Docker: Up and Running: Shipping Reliable Containers in Production* by *Sean P. Kane* and *Karl Matthias*

- *Continuous Delivery with Docker and Jenkins: Delivering software at scale* by *Rafal Leszko*

- *Docker in Action* by *Jeff Nickoloff* and *Stephen Kuenzli*

Answers

1. The `docker run` command.

2. The `-v`, or `--volume`, flag is used to attach a host machine path to a container path during execution.

3. `docker ps` or `docker container ls`.

14

Publishing Your Go Binary as a Homebrew Formula with GoReleaser

In this chapter, we'll look at GoReleaser and GitHub Actions and how they can be used in tandem to automate the release of a Go binary as a Homebrew formula. First, we'll look at **GoReleaser**, a popular open source tool that streamlines the creation, testing, and distribution of Go binaries. We'll look at its various configurations and options as well as how it works with GitHub Actions.

Following that, we'll look at **GitHub Actions**, a CI/CD platform that lets you automate software development workflows and integrate with other tools such as GoReleaser. We'll look at how to use it to ensure consistent and reliable builds, tests, and deployments.

After we've mastered both tools, we'll concentrate on triggering releases, creating a Homebrew tap, and integrating with Homebrew for simple installation and testing. **Homebrew**, a popular package manager for macOS, can be used for easy installation and management of your CLI application. Releasing your software to Homebrew not only simplifies the installation process for macOS users, but also gives you access to a wider audience. You can reach a sizable community of macOS developers and consumers that prefer using a package manager for program installation, such as Homebrew. Users can quickly find and install your software with just one command, increasing its usability and accessibility. This can help you reach a larger audience than you otherwise would and boost the visibility, usage, and adoption of your program.

By the end of this chapter, you'll have a firm grasp of how to combine GoReleaser and GitHub Actions to create an automated and efficient release process, and that includes publishing to Homebrew. With this knowledge, you will be able to tailor your own workflow to your specific requirements. The following topics will be covered:

- GoReleaser workflow

- Trigger release

- Installing with Homebrew and Testing

Technical requirements

For this chapter, you will need to do the following:

- You can also find the code examples on GitHub at `https://github.com/ PacktPublishing/Building-Modern-CLI-Applications-in-Go/tree/ main/Chapter14/audiofile`

- A GitHub account

- Install the GoReleaser tool at `https://goreleaser.com/install/`

GoReleaser workflow

Releasing software may be a lengthy and challenging process, particularly for projects with several dependencies and platforms. In addition to saving time, automating the release process lowers the possibility of human error and guarantees reliable and effective releases. GoReleaser is a popular choice for automating the release process for Go developers. However, there are also other choices, such as CircleCI, GitLab CI, and GitHub Actions, each of which has particular advantages and features. In this section, we'll examine the advantages of automating the release procedure and look in more detail at a few of these choices, specifically GoReleaser and GitHub Actions.

When compared to alternatives, GoReleaser stands out for the following reasons:

- **Easy to use**: Setup is simple and intuitive, making it easy for developers to get started with release automation. Their CLI quickly initializes a repository with a default configuration that can often work out of the box.

- **Platform support**: A variety of operating systems, including the major OSs, and cloud services are supported.

- **Customization at each step**: Programmers have the ability to customize at every step of the release process including building, testing, and publishing to various platforms.

- **Release artifacts**: A wide range of release artifacts may be produced, including Debian packages, Docker images, and binary files.

- **Versatile**: Combining with CI/CD pipelines, such as GitHub Actions, enables developers to fully automate their release procedure.

- **Open source**: Programmers may access the source code of the GoReleaser project and alter it to suit their own needs.

- **Community support**: GoReleaser offers a sizable and active user base, making it simple for developers to contribute to the project and find answers to their questions.

While there are many benefits to using GoReleaser, there are, however, a few reasons to not use GoReleaser for your project that you may need to consider:

- **Reliance on Github**: This may not be ideal if you prefer to use different tools or workflows.

- **Specific platform requirements**: While GoReleaser supports many popular OSs or cloud providers, you may require a platform that is not supported.

- **Complex release requirements**: While there is customization allowed in every step, there's a possibility GoReleaser may not be flexible enough to serve your specific level of complexity.

In conclusion, while there are other options out there, choose a tool that works for your specific use case. We do feel like GoReleaser is a great tool to use for the audiofile CLI use case, so let's continue.

Defining the workflow

Having analyzed the pros and cons of using GoReleaser, let's dive into its workflow by first sketching the overall process and then delving into each stage in greater detail:

1. Configure your project to use GoReleaser.
2. Configure GitHub Actions.
3. Set up your GitHub repositories.
4. Set up your GitHub token for Actions.
5. Tag and push the code.

It might make more sense to view this with a visual, so this is what we are trying to accomplish:

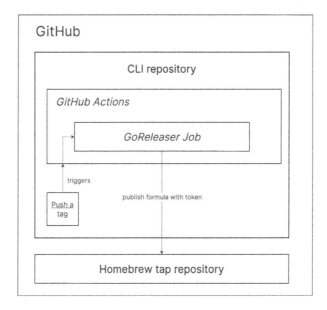

Figure 14.1 – Flow of the release process using GitHub Actions and GoReleaser

Let's dive deeper and gain a more thorough understanding of each step involved in the workflow now that we have a general idea of how it works; we will explore using GoReleaser with GitHub Actions and learn how to automate your own releases.

Configuring your project to use GoReleaser

With the GoReleaser tool installed, you can now initialize your CLI repository. In this case, we'll initialize the root of the audiofile CLI project repository by executing the following command:

```
goreleaser init
```

You should notice that a new file has been generated: `goreleaser.yml`. Before even inspecting the file, we can run a local-only release to confirm that there are no issues with the configuration by executing the following command:

```
goreleaser release --snapshot --clean
```

The output from the command gives you a clear idea of all the steps involved in the release. We will walk through these steps in the next section, *Trigger release*. At the very end of the output, you should see a message indicating a successful release, similar to this:

- `release succeeded after 10s`

While the default configuration succeeded, we'll need to look deeper into the configuration settings and modify and add appropriately to customize our release process. First, let's look at the default `.goreleaser.yaml` file together and break it down.

Global hooks

At the very top of the file, we see some default global hooks. **Hooks** allow you to run custom scripts. In GoReleaser, the `before` field allows you to specify which custom scripts to run before the release process begins:

```
before:
  hooks:
    - go mod tidy
    - go generate ./...
```

In the preceding example, we configured the automation to run the `go mod tidy` and `go generate ./...` commands prior to the release process. However, you might run scripts that perform tasks such as the following:

- Updating the version number in your project's code
- Generating a changelog
- Running automated tests to ensure that your code is working as expected
- Building your project and creating release artifacts
- Pushing changes to your version control system

The scripts you call from the `before` hooks section in GoReleaser can be written in any language, as long as they can be executed from the command line. For example, you might write scripts in Go, Python, Bash, or any other language that supports shell execution.

Builds and environment variables

Next, we see some default builds and some environment variables set. The `builds` field allows you to determine the combination of operating systems, defined by the `GOOS` field, architectures defined by the `GOARCH` field, and architecture mode, defined by the `GOARM` field. It also allows you to add additional fields, such as the `env` field, which allows you to set environment variables for the builds. Additional aspects that can be defined include the binary, flags, hook and more:

```
builds:
  - env:
      - CGO_ENABLED=0
    goos:
```

```
    - linux
    - windows
    - darwin
```

In the preceding example that exists within the default configuration, we defined the environment variable, `CGO_ENABLED`, to be set to `0`, and then configure the build process to generate binaries for the Linux, Windows, and Darwin operating systems.

> **Note**
>
> The `env` field can be set at a global level so that the environment variable is available at all stages of the release process, or it may be specified, such as in the previous case, within the build context alone.

The final configuration ends up having a few more modifications, such as specifying a few additional architectures, `amd64` and `arm64`, and removing `go generate ./...` before hook, which is unnecessary. Also, we've modified the `builds` field by setting build flags to use the `pro` and `dev` flags:

```
flags:
  - -tags=pro dev
```

While there are many other options you can set under the `builds` field, we will not cover them in this section. We encourage you to view the full list of customizations available at `https://goreleaser.com/customization/builds/`.

Archives

Next, we view some default `archives` settings. In GoReleaser, an **archive** is a compressed file that contains your application's binaries, `README`, and `LICENSE` files. The objective is to package your application's critical components into a single file, essentially making it easier to distribute and deploy. The default configuration sets the `archives` field to the following:

```
archives:
  - format: tar.gz
    name_template: >-
        {{ .ProjectName }}_
        {{- title .Os }}_
        {{- if eq .Arch "amd64" }}x86_64
        {{- else if eq .Arch "386" }}i386
        {{- else }}{{ .Arch }}{{ end }}
```

```
    {{- if .Arm }}v{{ .Arm }}{{ end }}
format_overrides:
- goos: windows
  format: zip
```

Within the preceding defaults for the `archives` field, you can see that the default archive format is `tar.gz` for all operating systems except for when GOOS is set to Windows. In that case, the archive format is `zip`. The name of the file is defined by a template. To learn more about the GoReleaser naming template, visit `https://goreleaser.com/customization/templates/` as there are many fields to customize the name of your archive. Let's at least review the keys used in the `naming_template` field:

- `.ProjectName` – the project name. If not set, GoReleaser will use the name of the directory containing the Go project. In our case, it is `audiofile`.

- `.Os` – GOOS value.

- `.Arch` – GOARCH value.

- `.Arm` – GOARM value.

Now that we understand what these template keys refer to, let's suppose that we generate an archive for our audiofile CLI project, for Linux, with an `amd64` architecture. The resulting name of the archive file would be `audiofile_Linux_x86x64.tar.gz`.

Checksum

GoReleaser automatically creates and includes a file called `project 1.0.0 checksums.txt` with the release package. **Checksums** allow your users to ensure that the files they've downloaded are complete and accurate. Similar to the `archives` field, you can use `naming_template` to generate the name of the `checksum` file. However, within our configuration, the default value for the `checksum` field is simply `checksums.txt`:

```
checksum:
  name_template: 'checksums.txt'
```

Defining a `checksum` file is important since it helps ensure the integrity of the data being distributed. The `checksum` file contains a one-of-a-kind code that can be used to verify that the downloaded files are identical to the original files. If the `checksum` file is not provided, the released files may be modified or corrupted throughout the download process. This can result in unpredictable behavior in your application and problems for your users. To avoid this, always provide a `checksum` file with your releases so that everyone knows they're getting the correct version of your product.

Snapshot

The `snapshot` field in the GoReleaser configuration file specifies whether a release is a "snapshot" or a stable release. A **snapshot** is a non-production version of a software project that is made available for testing and feedback.

The generated release artifacts will be marked as snapshots if the `snapshot` field is set to `true`. This means that the version number will be suffixed with `-SNAPSHOT`, and the release will not be published to any remote repository, such as GitHub Releases. If the `snapshot` field is set to `false` or is not supplied, the release is considered stable and is published normally. Like the previous two fields, `archives` and `checksum`, the `snapshot` field also has a `name_template` that can be used:

```
snapshot:
  name_template: "{{ incpatch .Version }}-next"
```

If not set, the default version, `0.0.1`, is set. Based on the previous template, the name of the snapshot will be `0.0.1-next`. `incpatch`, according to the GoReleaser documentation, increments the patch of the given reversion, with a side note that it will panic if it's not a semantic version. A **semantic version**, also known as **SemVer**, is a version numbering scheme that uses a format of `major.minor.patch` to convey the level of changes in a software release.

Changelog

The `changelog` field defines the path of your project's changelog file. A **changelog** file contains a list of all the changes, improvements, and bug fixes made to a software project, typically organized by version.

The aim is to record these changes so that users and developers can easily discover what's new in a specific release. The changelog also aids with debugging and support by documenting the development process. Let's look at the default configuration for the `changelog` field:

```
changelog:
  sort: asc
  filters:
    exclude:
      - '^docs:'
      - '^test:'
```

In the preceding block of the configuration, we defined the behavior of the changelog generation process. Using the `sort` field, we specify the order in which the entries of the changelog should be displayed, in this case, `asc`, for ascending. The `filters` field specifies, with the `exclude` subfield, a list of regular expressions matched against commits to be excluded. To view all the options available for the `changelog` field, visit `https://goreleaser.com/customization/changelog/`.

So, now that we've finished analyzing the default GoReleaser configuration, let's determine what we'd want to consider adding.

Release

The following code block in the GoReleaser configuration dictates that if there are any changes present within a Git repository, it automatically generates a pre-release. The pre-released version will have a version number that includes a pre-release suffix, such as `1.0.0-beta.1`:

```
release:
  prerelease: auto
```

This automated process provides a convenient way for developers to create early versions of their software for testing purposes. By utilizing pre-releases, they can quickly and effortlessly gather feedback on the latest changes and make any necessary modifications before releasing a final version to the public.

Universal binaries

Imagine having just one file that can work on multiple architectures of an operating system, such as an install that works on a macOS machine with either an M1 or Intel chip. That's what a **universal binary** is, also known as **fat binaries**. Instead of having separate binaries for different architectures, you'd have just one universal binary that can work on both. This makes it a lot more convenient for developers to spread their software across different platforms, and for users to just download a single file and run it on their system without having to worry about compatibility issues:

```
universal_binaries:
  - replace: true
```

We tell GoReleaser to use universal binaries by adding the `universal_binaries` field and setting the `replace` value to `true`.

Brews

The `brews` field allows developers to specify the details for creating and publishing Homebrew as part of their release process. Let's take a look at the following addition to our configuration for the audiofile CLI project:

```
brews:
  -
    name: audiofile
```

```
homepage: https://github.com/marianina8
tap:
  owner: marianina8
  name: homebrew-audiofile
commit_author:
  name: marianina8
```

Let's at least define what these fields are defining for the Homebrew creation and publishing process. A **tap repository** is a GitHub repository that contains one or more formula files, which define how to install a particular package on Homebrew. Note that the tap repository, although defined in the configuration, will be created in *Step 3*, *Setting up your GitHub repositories*:

- `name` – Defaults to the project name, audiofile.

- `homepage` – Your CLI application's homepage. This defaults to empty, but set it to your GitHub repository name.

- `tap` – Defines the GitHub/GitLab repository to publish the formula to. The `owner` field is the owner of the repository. The `name` field is the name of the repository.

- `commit_author` – This is the Git author that shows up when committing to the repository. It defaults to `goreleaserbot`, but in our case, we set it to our GitHub handle.

You can view all the available customizations available for the `brew` field at `https://goreleaser.com/customization/homebrew/`.

Onto the next step!

Configuring GitHub Actions

Within this section, we'll learn about GitHub Actions and how they can be integrated with the GoReleaser tool. First, **Github Actions**, as you may recall, is a CI/CD tool, but get ready for this, it also has an incredible feature that allows you to set off an execution of whatever code you like on your repository when a certain event occurs! You may already know this actually, but for those who now know, new doors of opportunity are opening. Let's discuss the main components of GitHub Actions:

- **Events**: Any GitHub event, such as pushing code, creating a new branch, opening a PR, a pull request, or commenting on an issue. Events trigger workflows.

- **Runners**: A runner is a process that starts executing a workflow when triggered by an event. There is a one-to-one relationship between a runner and a job.

- **Workflows**: Workflows are composed of a series of jobs that can run sequentially or in parallel. They are defined in the `./.github/workflows` directory.

- **Jobs**: A single job is a series of tasks. A task could be a script or another GitHub action.

- **Actions**: An action is a task. Some tasks may perform complex tasks such as publishing a Go package to Homebrew, or simple tasks, such as setting an environment variable.

The following diagram may help to indicate the relationship between the four major components of GitHub Actions:

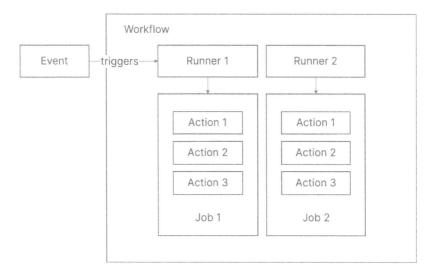

Figure 14.2 – Relationship between events, runners, workflows, jobs, and actions

Now that we've grasped the concept of GitHub Actions, let's see how we can take events, such as pushing tags, to trigger a GoReleaser job, which performs the complex task of publishing a Go package to Homebrew for us. First, we need to create the configuration file. From the root of the repository, do the following:

1. Create a folder called `.github`.

2. Create a subfolder inside of the `.github` folder, called `workflows`.

3. Create a `release.yml` file.

The GoReleaser website provides a default configuration for GitHub Actions on their website at `https://goreleaser.com/ci/actions/`. You may copy and paste this from their website to give you a good starting point. We'll make a few modifications, but before we do, let's walk through the default configuration together. Let's begin by discussing the fields that exist within the GitHub Actions `release.yml` file.

On

The `on` field within the Github Actions repository specifies the events that trigger the workflow. It could be a single event or multiple. Let's go through some of the events:

*   Push: The `push` field is used to tell the action to trigger a push. For example, this can be customized to specify pushing to a branch or tag. The syntax for this field is defined as follows:

    ```
    on.push.<branches|tags|branches-ignore|tags-ignore>.<paths|paths-ignore>
    ```

- Use the `branches` filter to include specific branch names and the `branches-ignore` filter to exclude certain branch names. Remember, don't use both `branches` and `branches-ignore` in the same workflow event.

- Use the `tags` filter for including specific tag names and `tags-ignore` for excluding certain tag names. Again, be careful, don't use both `tags` and `tags-ignore` in the same workflow! Apparently, the workflow won't run if that's the case.

- The `paths` and `paths-ignore` fields can be used to specify if code has changed within a particular path. The value of those fields may be set to glob patterns that use the `*` and `**` wildcard characters. The `paths` and `paths-ignore` filters allow you to control what paths are included or excluded from path pattern matching.

- Pull request: The `pull_request` field is used to tell the action to trigger a pull request. Like the previous field, we can specify the `branches` filter to include specific branch names or the `branches-ignore` filter to exclude branch names. Similarly, the `paths` and `paths-ignore` fields may also be set. The `branches` and `branches-ignore` fields also accept glob patterns.

- event_name: The `event_name` field defines the type of activity that will trigger a workflow to be executed. Within GitHub, there are events that can be triggered from more than one activity. The syntax of the full defined event that includes this field is defined as follows:

```
on.<event_name>.types
```

There is quite a long list of available events that can be used, including the two that we covered earlier, push and `pull_request`, but also `check_run`, `label`, `release`, and many more.

There's a lot more that can be done with GitHub Actions, so to see the full list of options to customize the on field, visit `https://docs.github.com/en/actions/using-workflows/workflow-syntax-for-github-actions`.

Now that we have a good understanding of GitHub Actions, let's look at the default configuration and see what it has set for the on field:

```
on:
  push:
    # run only against tags
    tags:
      - '*'
```

Perfect! This is pretty much exactly what we need. The preceding block of code specifies running the workflow triggered by tags being pushed.

Permissions

The permissions field is used to define the level of access that the GitHub Actions workflow has to various resources within your GitHub repository. Essentially, it helps you control what your workflow can and can't do within your repository. Let's take a look at the default configuration for the permissions field:

```
permissions:
  contents: write
  # packages: write
  # issues: write
```

The last two lines are commented out, but we can still discuss them. Within the preceding code, there are three permission types specified: contents, packages, and issues. Since these permissions are all set to write, but the latter two are commented out, then we limit the workflow permissions to contents: write to the repository. Based on the documentation on GoReleaser, the contents:write permission is required in order to upload archives as GitHub releases or to publish to Homebrew.

If you want to push Docker images to GitHub, you'll need to enable the packages: write permission. If you use milestone closing capacity, you'll need the issues: write permission enabled.

Jobs

The jobs field defines the individual tasks that make up your workflow. It is basically the blueprint of the workflow, defining each job and specifying in which order they will be executed. Let's take a look at the default value set within our configuration file:

```
jobs:
  goreleaser:
    runs-on: ubuntu-latest
    steps:
      - uses: actions/checkout@v3
        with:
          fetch-depth: 0
      - run: git fetch --force --tags
      - uses: actions/setup-go@v3
        with:
          go-version: '>=1.20.0'
```

```
        cache: true
  - uses: goreleaser/goreleaser-action@v4
    with:
      distribution: goreleaser
      version: latest
      args: release --clean
    env:
      GITHUB_TOKEN: ${{ secrets.GITHUB_TOKEN }}
```

Now, let's clarify the preceding code. There is only one job, named `goreleaser`, defined under the `jobs` field. The `goreleaser` job has the following steps defined in order:

1. **Checkout code** – the first step uses the `actions/checkout@v3` action to check out the code from your GitHub repository. The `fetch-depth` parameter is set to 0, which ensures that all branches and tags are fetched from the repository.

2. **Fetch git tags** – the second step runs the `git fetch --force --tags` command, which fetches all tags from the Git repository.

3. **Set up Go environment** – the third step uses the `actions/setup-go@v3` action to set up a Go environment. The `go-version` parameter is set to `>=1.20.0`, which specifies the minimum version of Go required for this job. The `cache` parameter is set to `true`, which tells GitHub Actions to cache the Go environment, speeding up subsequent runs of this job.

4. **Release with GoReleaser** – the final step uses the `goreleaser/goreleaser-action@ v4` action to release the code with GoReleaser. The `distribution` parameter is set to `goreleaser`, which specifies the type of distribution to be used. The `version` parameter is set to `latest`, which specifies the latest version of GoReleaser to be used. The `args` parameter is set to `release --clean`, which specifies the command-line arguments to be passed to GoReleaser when the release is performed.

The only modification that we'll need to do to the default configuration is to modify the `with. version` field for the `goreleaser/goreleaser-action` step. Currently, the default value is set to `latest`. Let's replace it with `${{ env.GITHUB_REF_NAME }}`. The environment variable, `env.GITHUB_REF_NAME`, is automatically set by GitHub and represents the branch or tag name for the current Git reference.

Final note, at the bottom of the configuration file, the environment variables are set to be used when `goreleaser` runs. `secrets.GITHUB_TOKEN` must be replaced with `secrets.PUBLISHER_ TOKEN`. This token will be used when publishing to our other repository, the Homebrew tap repository. We've completed the configuration of our GitHub Actions, so now we can move on to the next step.

Setting up your GitHub repositories

If you've been following along with the audiofile CLI repository, then the repository already exists on GitHub. However, if you are creating your own CLI application in tandem, now is the time to make sure the repository exists on GitHub.

Besides pushing your CLI application's repository to GitHub, we'll also need to create the Homebrew tap repository that was defined earlier in the GoReleaser configuration file. A Homebrew **tap repository** is a second location, in our case a repository hosted on GitHub, that is separate from the official Homebrew repository, which hosts additional formula for Homebrew to install on macOS. Taps make it easier for users to install new applications, that do not exist in the `homebrew/core` repository, onto their computers.

Let's follow the steps to creating a new Homebrew tap repository:

1. Sign in to GitHub `https://github.com`.
2. Click the **New repository** button from your GitHub dashboard.
3. Enter the repository details. In our example, enter the name, homebrew-audiofile, that matches what we set within the GoReleaser configuration. Make sure that the repository is set to `Public` as well.
4. Create the repository by clicking on the **Create repository** button.
5. Clone the repository to your local machine.

There's no reason to add any files at this point. The GoReleaser tool, once we run the release process, will push the formula to this repository, but first, we need to create a token to use.

Setting up your GitHub Token for Actions

In order to make the GoReleaser and GitHub Actions workflow work, we need to create a GitHub token and Actions secret.

To create a GitHub Token, click on your user menu and select the **Settings** option:

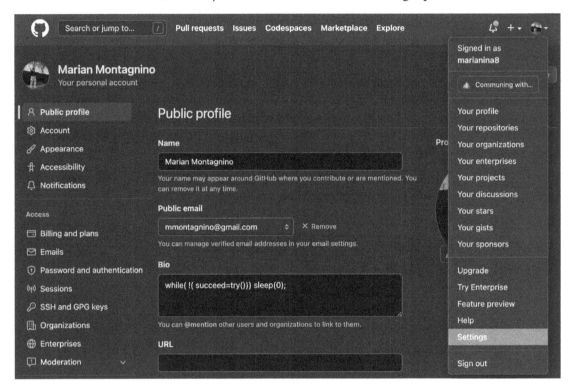

Figure 14.3 – User menu with the Settings option selected

Once you are on the **Settings** page, scroll down the menu to see the last option, **Developer Settings**. When you select **Developer Settings**, you should now see the **Personal access tokens** option in the menu on the left side.

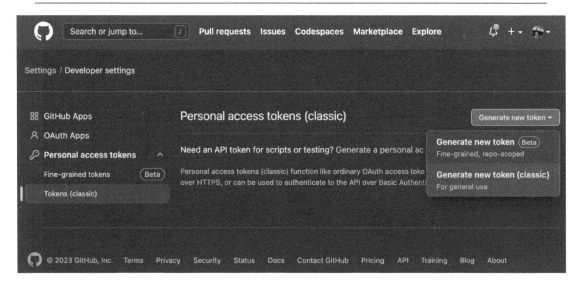

Figure 14.4 – The Developer Settings page with Generate new token options

Click on the **Generate new token** button. You might need to authenticate again if you have a two-factor authentication setup, but then you should be routed to the **New personal access token (classic)** page. From this page, take the following steps to create your GitHub token:

1. Enter a value for the **Note** field, a description of what the token will be used for. Let's put in `audiofile` since this will be used for the audiofile CLI project.

2. In the **Select scopes** section, select **repo**. This will give it permission to run actions against your repositories. Then, scroll to the bottom and click the **Generate token** button.

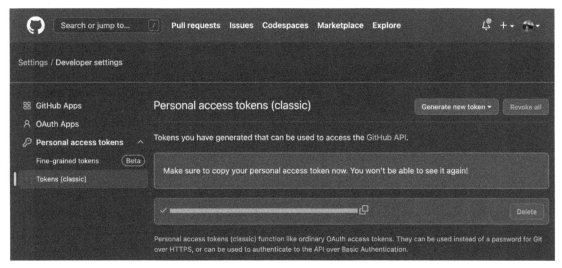

Figure 14.5 – The Personal access tokens page once the token has been generated

3. Copy the generated token (which has been blocked out within the preceding screenshot).

4. Go back to your CLI repository; in our case, we went back to the `https://github.com/marianina8/audiofile`.

5. Click on **Settings**.

6. From the menu on the left side, click on **Secrets and Variables**, which expands to show more options. Click on the **Actions** option.

7. Click on **New repository secret**, which is in the top right corner of the screen.

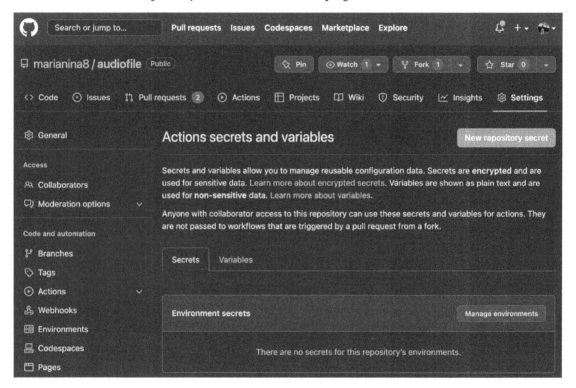

Figure 14.6 – Actions secrets and variables page

8. From the **Actions secrets/New secret** page, enter the **Name** of your secret. This should match what we entered within the GoReleaser configuration. If you recall, the `env.GITHUB_TOKEN` value in the template was set to `secrets.PUBLISHER_TOKEN`. Enter the `PUBLISHER_TOKEN` value into the **Name** field.

9. Paste the secret you copied in *step 3* into the **Secret** field.

10. Click the **Add secret** button.

11. Confirm that the secret now exists in your **Actions secrets and variables** page.

Figure 14.7 – Repository secrets showing newly created PUBLISHER_TOKEN

Now that the publisher token is set, let's move on to the final step.

Trigger release

Now that the configuration files have been set up for GoReleaser and GitHub Actions, and the publisher tokens to give access to make changes to the repositories are also created and shared, we are ready to trigger a release with the next step of the workflow: tag and push the code. Before we do so, let's take a step back and discuss what happens when you trigger the goReleaser job:

- **Preparation**: GoReleaser checks the configuration files, validates the environment, and sets up the necessary environment variables

- **Building**: Builds the Go binary and compiles it for multiple platforms (such as Windows, Linux, and macOS)

- **Versioning**: Generates a new version number based on the existing version and the user's configuration

- **Creating the release artifacts**: Generates the release artifacts, such as tarballs, deb/rpm packages, and zip files for each platform

- **Creating a Git tag**: Creates a new Git tag for the release, which is used to reference the release in the future

- **Uploading the artifacts**: Uploads the generated release artifacts to the specified locations, such as a GitHub release, a file server, or a cloud storage service

- **Updating Homebrew formulas**: If you are using Homebrew, it will update the Homebrew formulas to reflect the new release

- **Notifying stakeholders**: If set up to do so, GoReleaser can notify stakeholders about the new release through various channels, such as email, Slack, or webhooks

Note that the previous steps may vary based on the specific configuration and plugins used with GoReleaser. Moving on, let's trigger it with the push of a tag.

Tag and push the code

At this point, make sure all the code changes you have been pushed to the remote repository for your CLI project:

1. Tag your CLI with the appropriate version. For our CLI project, within the audiofile repository, we run the following Git command:

    ```
    git tag -a v0.1 -m "Initial deploy"
    ```

2. Now push the tag to the repository. This should trigger the GitHub Actions to take place:

    ```
    git push origin v0.1
    ```

3. Visit the CLI repository, and you'll now notice a yellow dot appear at the top of the file listing:

Figure 14.8 – Repository showing yellow dot

4. Click on the yellow dot, and a popup will appear. To view the details of the GoReleaser process, click on the **Details** link:

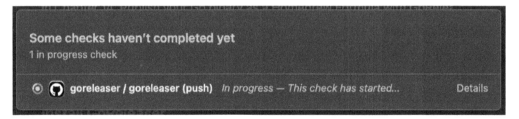

Figure 14.9 – Details popup of goReleaser process

5. Clicking on the **Details** link will take you to a page where you can watch the GoReleaser workflow progressing through each task:

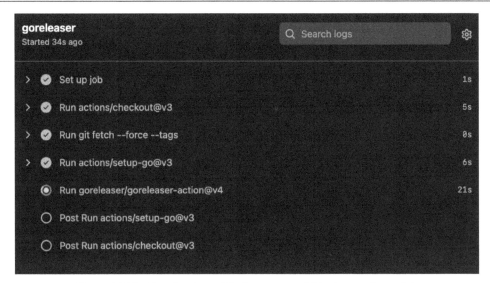

Figure 14.10 – List of tasks and their progress within the goreleaser job

6. Once it successfully completes, from the CLI repository, click on the tag listed under the **Releases** section on the right-hand side of the page. From there, you'll see the changelog and list of **Assets**:

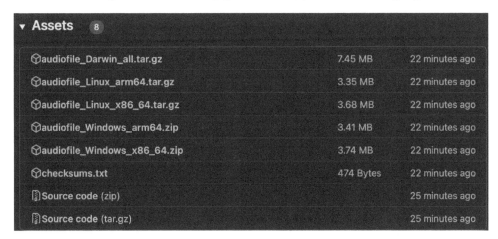

Figure 14.11 – List of assets generated from the goreleaser job

Looks like all the builds were successfully generated and archived and are available as assets on the **Releases** page. What if it can be installed successfully with Homebrew? For the final confirmation, let's jump to the next section.

Installing with Homebrew and Testing

Since the GoReleaser job ran successfully, we should be able to install the CLI application with Homebrew. Let's begin by telling Homebrew to tap the repository we've created for the formula:

```
brew tap marianina8/audiofile
```

You should see the following output generated from the previous command:

```
==> Tapping marianina8/audiofile
Cloning into '/opt/homebrew/Library/Taps/marianina8/homebrew-
audiofile'...
remote: Enumerating objects: 6, done.
remote: Counting objects: 100% (6/6), done.
remote: Compressing objects: 100% (4/4), done.
remote: Total 6 (delta 0), reused 3 (delta 0), pack-reused 0
Receiving objects: 100% (6/6), done.
Tapped 1 formula (13 files, 6.3KB).
```

As we know, tapping the repository adds to the list of Homebrew formulas to install. Next, let's try installing the audiofile CLI:

```
brew install marianina8/audiofile/audiofile
```

You should see the following output generated for the installation of the application:

```
==> Fetching marianina8/audiofile/audiofile
==> Downloading https://github.com/marianina8/audiofile/
releases/download/v0.2/audiofile_Darwin_all.tar.gz
==> Downloading from https://objects.githubusercontent.com/
github-production-release-asset-2e65be/483881004/ccc2302f-a4a5-
454a
############################################################
######### 100.0%
==> Installing audiofile from marianina8/audiofile
🍺  /opt/homebrew/Cellar/audiofile/0.2: 4 files, 19.2MB, built
in 3 seconds
==> Running `brew cleanup audiofile`...
Disable this behaviour by setting HOMEBREW_NO_INSTALL_CLEANUP.
Hide these hints with HOMEBREW_NO_ENV_HINTS (see `man brew`).
```

Now, for the final test, let's run the `audiofile` command and see the output:

```
● mmontagnino@Marians-MacBook-Pro audiofile % audiofile
  A command line interface allows you to interact with the Audiofile service.
  Basic commands include: get, list, and upload.

  Usage:
    audiofile [command]

  Available Commands:
    api         Start or stop the API required by the CLI
    bug         Submit a bug
    completion  Generate the autocompletion script for the specified shell
    delete      Delete audiofile by id
    get         Get audio metadata
    help        Help about any command
    list        List all audio files
    play        Play audio file by id
    player      Launch player dashboard
    search      Command to search for audiofiles by string
    upload      Upload an audio file
    version     Version of audiofile CLI

  Flags:
    -h, --help      help for audiofile
    -t, --toggle    Help message for toggle
    -v, --verbose   verbose

  Use "audiofile [command] --help" for more information about a command.
```

Figure 14.12 – Output of the audiofile command installed by Homebrew

Let's now try some of the commands; first, let's start the API in one terminal window:

```
mmontagnino@Marians-MacBook-Pro audiofile % audiofile api
Starting API at http://localhost:8000
Press Ctrl-C to stop.
```

In another terminal, let's run the player by calling the following:

```
audiofile player
```

You should see the following:

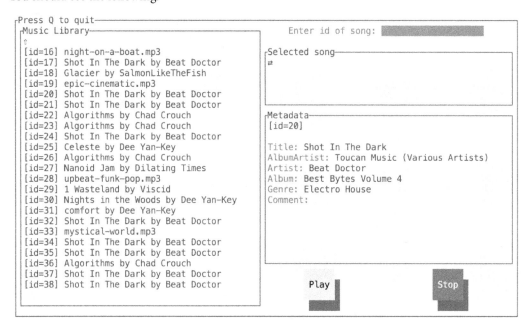

Figure 14.13 – The audiofile player

We've been able to install with the Homebrew package manager and play around with the audiofile to know that it works well. This concludes our chapter on publishing your Go binary as a Homebrew formula with GoReleaser. While Homebrew is just one package manager, you can follow a similar process for **GoFish**, a cross-platform systems package manager that allows users to easily install applications on Linux and Windows. Combined, you'll be able to expand your user base and make it easy for your users to install and update your CLI application.

Summary

In this chapter, we took a closer look at how GoReleaser and GitHub Actions can work together to make releasing a CLI application a breeze. First, we got to know GoReleaser, a handy tool that makes building, testing, and deploying Go binary packages a snap. We went over the default configuration file and also talked about some simple tweaks you can make to fit your needs. Then, we explored GitHub Actions and how to integrate them with GoReleaser.

By the end of this chapter, we had a good understanding of how to use these tools to create a seamless and efficient release process, complete with publishing on Homebrew. Releasing through Homebrew opens up the possibility of reaching more users who prefer to use package managers.

Questions

1. When are `before` hooks run? Are there `after` hooks?

2. What is the `PUBLISHER_TOKEN` GitHub token used for?

3. Can you trigger a GitHub Action workflow on a pull request?

Further reading

- GoReleaser documentation can be found at `https://goreleaser.com/`

- GitHub Actions documentation can be found at `https://docs.github.com/en/ actions`

- Homebrew documentation can be found at `https://docs.brew.sh/`

Answers

1. The `before` hooks field specifies scripts that are run before the release process. Yes, although not discussed in this chapter, there are `after` hooks, too!

2. The `PUBLISHER_TOKEN` GitHub token is set as an environment variable on the goreleaser job in the `release.yml` file that defines the GitHub Actions release workflow. The token is configured within GitHub to give repository access to GitHub Actions, allowing the `goreleaser` job to publish the Homebrew formula to the `homebrew-audiofile` tap repository.

3. Yes, among many other triggers described in this chapter, pull requests can also trigger GitHub Action workflows.

Index

www.packtpub.com

Subscribe to our online digital library for full access to over 7,000 books and videos, as well as industry leading tools to help you plan your personal development and advance your career. For more information, please visit our website.

Why subscribe?

- Spend less time learning and more time coding with practical eBooks and Videos from over 4,000 industry professionals

- Improve your learning with Skill Plans built especially for you

- Get a free eBook or video every month

- Fully searchable for easy access to vital information

- Copy and paste, print, and bookmark content

Did you know that Packt offers eBook versions of every book published, with PDF and ePub files available? You can upgrade to the eBook version at www.packtpub.com and as a print book customer, you are entitled to a discount on the eBook copy. Get in touch with us at customercare@packtpub.com for more details.

At www.packtpub.com, you can also read a collection of free technical articles, sign up for a range of free newsletters, and receive exclusive discounts and offers on Packt books and eBooks.

Other Books You May Enjoy

If you enjoyed this book, you may be interested in these other books by Packt:

Event-Driven Architecture in Golang

Michael Stack

ISBN: 9781803238012

- Understand different event-driven patterns and best practices
- Plan and design your software architecture with ease
- Track changes and updates effectively using event sourcing
- Test and deploy your sample software application with ease
- Monitor and improve the performance of your software architecture

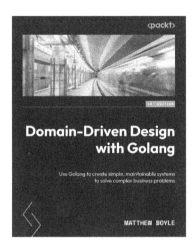

Domain-Driven Design with Golang

Matthew Boyle

ISBN: 9781804613450

- Get to grips with domains and the evolution of Domain-driven design
- Work with stakeholders to manage complex business needs
- Gain a clear understanding of bounded context, services, and value objects
- Get up and running with aggregates, factories, repositories, and services
- Find out how to apply DDD to monolithic applications and microservices
- Discover how to implement DDD patterns on distributed systems
- Understand how Test-driven development and Behavior-driven development can work with DDD

Packt is searching for authors like you

If you're interested in becoming an author for Packt, please visit `authors.packtpub.com` and apply today. We have worked with thousands of developers and tech professionals, just like you, to help them share their insight with the global tech community. You can make a general application, apply for a specific hot topic that we are recruiting an author for, or submit your own idea.

Share your thoughts

Now you've finished *Building Modern CLI Applications in Go*, we'd love to hear your thoughts! Scan the QR code below to go straight to the Amazon review page for this book and share your feedback or leave a review on the site that you purchased it from.

https://packt.link/r/1804611654

Your review is important to us and the tech community and will help us make sure we're delivering excellent quality content.

Download a free PDF copy of this book

Thanks for purchasing this book!

Do you like to read on the go but are unable to carry your print books everywhere?

Is your eBook purchase not compatible with the device of your choice?

Don't worry, now with every Packt book you get a DRM-free PDF version of that book at no cost.

Read anywhere, any place, on any device. Search, copy, and paste code from your favorite technical books directly into your application.

The perks don't stop there, you can get exclusive access to discounts, newsletters, and great free content in your inbox daily

Follow these simple steps to get the benefits:

1. Scan the QR code or visit the link below

https://packt.link/free-ebook/9781804611654

2. Submit your proof of purchase

3. That's it! We'll send your free PDF and other benefits to your email directly

www.ingramcontent.com/pod-product-compliance
Lightning Source LLC
Chambersburg PA
CBHW062034050326
40690CB00016B/2939